Lecture Notes in Networks and Systems

Volume 180

The series “Lecture Notes in Networks and Systems” publishes the latest developments in Networks and Systems—quickly, informally and with high quality. Original research reported in proceedings and post-proceedings represents the core of LNNS.

Volumes published in LNNS embrace all aspects and subfields of, as well as new challenges in, Networks and Systems.

The series contains proceedings and edited volumes in systems and networks, spanning the areas of Cyber-Physical Systems, Autonomous Systems, Sensor Networks, Control Systems, Energy Systems, Automotive Systems, Biological Systems, Vehicular Networking and Connected Vehicles, Aerospace Systems, Automation, Manufacturing, Smart Grids, Nonlinear Systems, Power Systems, Robotics, Social Systems, Economic Systems and other. Of particular value to both the contributors and the readership are the short publication timeframe and the world-wide distribution and exposure which enable both a wide and rapid dissemination of research output.

The series covers the theory, applications, and perspectives on the state of the art and future developments relevant to systems and networks, decision making, control, complex processes and related areas, as embedded in the fields of interdisciplinary and applied sciences, engineering, computer science, physics, economics, social, and life sciences, as well as the paradigms and methodologies behind them.

Indexed by SCOPUS, INSPEC, WTI Frankfurt eG, zbMATH, SCImago.

All books published in the series are submitted for consideration in Web of Science.

More information about this series at http://www.springer.com/series/15179

Bogdan Ghita · Stavros Shiaeles
Editors

Selected Papers from the 12th International Networking Conference

INC 2020

Springer

Editors
Bogdan Ghita
School of Engineering, Computing
and Mathematics
University of Plymouth
Plymouth, UK

Stavros Shiaeles
School of Computing
University of Portsmouth
Portsmouth, UK

ISSN 2367-3370 ISSN 2367-3389 (electronic)
Lecture Notes in Networks and Systems
ISBN 978-3-030-64757-5 ISBN 978-3-030-64758-2 (eBook)
https://doi.org/10.1007/978-3-030-64758-2

This Springer imprint is published by the registered company Springer Nature Switzerland AG
The registered company address is: Gewerbestrasse 11, 6330 Cham, Switzerland

Message from Conference Chairs

It is our great honor and pleasure to welcome you to the 12th International Network Conference 2020 (INC2020), hosted virtually on the 21st of September 2020. The aim of the INC2020 conference is to establish an international forum for engineers and scientists to present their ideas and advancements in the field of networking and the related areas.

Networking represents the underlying core of current IT systems, providing the necessary communication support for complex infrastructures. Recent years have witnessed a number of novel concepts moving from theory to large-scale implementations, such as software-defined networking, network function virtualization, 5G, smart environments, and IoT. These concepts change fundamentally the paradigms used in traditional networking, with a number of areas such as network routing and system or business security having to adjust or redesign to accommodate them. While the benefits are clear, through the advent of new applications, use cases, improved user interaction and experience, they also introduce new challenges for generic network architectures, mobility, security, and traffic engineering. Through its topics, the INC2020 conference addresses the comprehensive nature of these technologies and the associated emerging challenges by discussing the approaches, tools, solutions, and advancements in these areas.

We are delighted to announce that we received many high-quality submissions that underwent a rigorous peer-review process on the basis of their significance, novelty, technical quality, and practical impact. A total of 16 high-quality research papers have been selected for presentation with topics focusing on malware analysis, encryption, biometric authentication and profiling, wireless secure transmission and attacks, SDN, TCP performance, IoT profiling and protection, cyber-range connectivity, traffic monitoring, VANET security, and WSN infrastructure deployment.

There is always a great team behind a successful event. We would like to take this chance to thank the organizing committee of the INC2020 conference for their effort and support toward the success of the conference. We want to express our sincere gratitude to all the authors, participants, technical programme committee members, and many others who greatly contributed to this conference. We sincerely hope you will find the INC2020 conference quite informative and interesting for your research and professional activities.

Bogdan Ghita
Stavros Shiaeles
Chairs of INC2020

Contents

Security

Malware Behavior Through Network Trace Analysis

Xiyue Deng(✉) and Jelena Mirkovic

Information Sciences Institute, University of Southern California, Los Angeles, USA
{xiyueden,mirkovic}@isi.edu

Abstract. Malware continues to be a major threat to information security. To avoid being detected and analyzed, modern malware is continuously improving its stealthiness. A high number of unique malware samples detected daily suggests a likely high degree of code reuse and obfuscation to avoid detection. Traditional malware detection techniques relying on binary code signatures are greatly hindered by encryption, packing, code polymorphism, and similar other obfuscation techniques. Although obfuscation greatly changes a malware's binary, its functionalities remain intact.

We propose to study malware's network behavior during its execution, to understand the malware's functionality. While malware may transform its code to evade analysis, we contend that its key network behaviors must endure through the transformations to achieve the malware's ultimate purpose, such as sending victim information, scanning for vulnerable hosts, etc. While live malware analysis is risky, we leverage the Fantasm platform on the DeterLab testbed to perform it safely and effectively. Based on observed network traffic we propose an encoding of malware samples. This encoding can help us classify malware flows and samples, identify code reuse and genealogy, and develop behavioral signatures for malware defense. We apply our approach to more than 8,000 diverse samples from the Georgia Tech Apiary project. We find that over 60% of malware is multi-purposed (e.g. downloading new payload and uploading user data). We also illustrate how our encoding and malware flow clustering can be used to identify behavioral signatures for malware defense.

Keywords: Malware · Network · Polymorphic · Genealogy

1 Introduction

The Internet is facing increasing threats due to the proliferation of malware. A study by Kaspersky Lab suggests an estimated daily increase of over 360,000 samples in the wild in 2017. [11]. Such high malware production rate suggests that new malware may be created by transforming existing code to evade signature detection.

Malware designers have invested enormous efforts to change their binaries to avoid detection and analysis by defenders. From simple instruction transformation, such techniques have evolved through junk code injection, code obfuscation,

B. Ghita and S. Shiaeles (Eds.): INC 2020, LNNS 180, pp. 3–18, 2021.
https://doi.org/10.1007/978-3-030-64758-2_1

to polymorphic and metamorphic engines that can transform malware code into millions of variants [19]. Such obfuscation techniques have greatly undermined traditional signature-based malware detection methods, and they create an enormous workload for code analysis. Yet much of the new malware variants could simply be old malware in a new package, or new malware assembled from pieces of old malware. Clearly, as new malware samples emerge, code analysis and signature-based filtering cannot keep pace.

Besides static analysis, researchers have used dynamic analysis to overcome code obfuscation. Dynamic analysis includes tracking disk changes, analyzing dynamic call graphs, as well as monitoring malware execution using debuggers and virtual machines. However, modern malware is often equipped with anti-debugging and anti-virtualization capabilities [20,21], making dynamic code analysis in a controlled environment difficult.

To complement contemporary static and dynamic *code* analysis, we propose to study malware behavior by observing and interpreting its *network activity*. Much of today's malware relies on the network connectivity to achieve its purpose, such as sending reports to the malware author, joining the botnet, sending spam and phishing emails, etc. We hypothesize that it would be difficult for malware to significantly alter its network behavior and still achieve its purpose. Studying network behavior thus may offer an opportunity to both understand what malware is trying to do in an analysis environment, and to develop behavior or network traffic signatures useful for malware defense.

Live malware analysis, with unrestricted network access, is risky, because malware may inflict damage to other Internet hosts during analysis, and analysts would become unwitting accomplices. We leverage the Fantasm system for safe and productive live malware analysis [6] to minimize this risk. The system enables us to emulate unrestricted connectivity from malware's point of view, while it protects the Internet from unwanted traffic.

Just having network traffic records of a malware run is not enough to understand a malware's purpose, because such data is very rich and unstructured. To overcome this challenge, we develop an *embedding* for a malware sample – a set of flow-record features observed during the analysis. We then perform a two-pronged analysis. First, we devise patterns over the embeddings that help us detect presence and features of high-level network behaviors, such as file download, e-mail sending, scanning, etc. We use a set of these high-level behaviors to infer a malware sample's purpose. We note that malware could have more than one purpose. For example, it could both scan for vulnerable hosts and send unsolicited emails. Thus a malware sample could have more than one label. Second, we use the embeddings to identify malware samples that have the same or very similar network behaviors, and study code reuse and malware genealogy.

We perform our analysis on 8,172 malware samples, randomly selected from the Georgia Tech Apiary project [1]. 63.9% of the samples exhibit more than one behavior, which speaks to the multi-purpose nature of contemporary malware. We further cluster malware flows based on their embeddings and identify groups that have similar or even identical behaviors. We find that only 14 clusters are responsible for over 80% of data-carrying flows, which appear in about 70% of malware samples. Apart from these flows, the samples exhibit diverse other network behaviors, ranging from repeatedly querying `google.com`, to uploading private user information, to utilizing Bittorrent tracker to scan for potential victims. We then show how our clusters can be used to devise behavioral signatures for malware defenses.

2 Background and Related Work

Malware analysis has received continuous increase of research interests [3,17,24]. In this section, we discuss contemporary malware analysis methods and the rationale behind our approach. We also survey related work.

2.1 Static Analysis

A common approach to malware detection is analyzing binaries for *code signatures* – sequences of binary code that are present in malware and are not common in benign binaries. Various static analysis methods like CTPL [12], Generic Virus Scanner [5], etc. have been used to analyze binary code of malware without running it, and identify portions that could be used for signature generation. The identified portion is then synthesized to become the signature of this kind of malware.

Signature-based malware detection has been the most widely used method and has been quite successful. However, malware designers have also been working on countermeasures over the years to undermine such techniques. From junk code generation, malware encryption and oligomorphism, to polymorphic and metamorphic malware [19], such techniques have evolved significantly. Our research complements signature-based detection by identifying common network-level behaviors of malware. These behaviors can be used to develop behavioral signatures of malware, which can be used to detect malware running on compromised hosts. In other words, signature-based detection can *prevent* malware infections, and *behavior signatures* can detect infections that bypass signature-based detection.

2.2 Dynamic Analysis on Host

Dynamic analysis builds signatures of a malware's interaction with its host. Such signatures may include memory access and file access patterns, as well as system call patterns. Those patterns that are likely to be present in malware but not in benign binaries can be used to develop a behavioral signature for malware detection [7].

Dynamic analysis complements static analysis and can overcome malware code obfuscation. Willems et al. [26] proposed CWSandbox that combines static and dynamic techniques for analyzing malware on a contained host. Guarnieri et al. [8]'s Cuckoo sandbox follows similar concepts as CWSandbox and additionally provides a network packet sink. Both works utilize virtual machines to isolate the study environment and the host. However, stealthy malware has another set of techniques to evade dynamic analysis – it detects debuggers and virtual machines, which are often used to speed up and facilitate dynamic analysis, and modifies its behavior. This leads to incorrect or unusable signatures of stealthy malware. Chen et al. [4] found that 39.9% and 2.7% of 6,222 malware samples exhibit anti-debugging and anti-virtualization behaviors respectively. In 2012, Branco et al. [2] analyzed 4 million samples and observed that 81.4% of them exhibited anti-virtualization behavior and 43.21% exhibited anti-debugging behavior.

Our work complements dynamic analysis, by providing another set of features, based on network behavior of malware, that can be used for detection. Our approach allows for feature collection using the network and does not require virtualization (binaries can be run on bare metal machines).

2.3 Dynamic Analysis of Network Behavior

There are a few efforts on analyzing the semantic of malware network behavior. Sandnet [18] provides a detailed, statistical analysis of malware network traffic, and surveys popular protocols employed by malware. Our work aims to understand popular network behavior patterns and the similarity in network behaviors between different malware samples.

Morales et al. [15] studied several network activities, selected using heuristics, which include: (1) NetBIOS name request, (2) failed network connections after DNS or rDNS queries, (3) ICMP-only activity with no replies or with error replies, (4) TCP activity followed by ICMP replies, etc. These activities can be used to detect the likely presence of malware. Morales et al. also report on the prevalence of those behaviors in contemporary malware. While their chosen activities are useful for malware detection, our high-level behavior analysis focuses on understanding malware purposes (e.g., spamming vs scanning). Our work thus complements the work by Morales et al.

Nari et al. [16] proposed an automated malware classification system also focusing on malware network behavior, which generates protocol flow dependency graph based on the IP address being contacted by malware. Our work improves this effort by systematizing detection of different malware behaviors using different network behavior patterns, and includes other information from network flows such as packet size, packet contents, etc.

Lever et al. [13] experimented with 26.8 million samples collected over five years and showed several findings including that dynamic analysis traces are susceptible to noise and should be carefully curated, Internet services are increasingly filled with potentially unwanted programs, as well as that network traffic provides the earliest indicator of infection. As a slight downside, the data used for this long period came from different sources which may not be collected in a well controlled environment and could result in noise that is hard to identify and

Table 1. Flow policies in Fantasm

Service or Protocol	Label
DNS, HTTP, HTTPS	Not risky
FTP, SMTP, ICMP_ECHO	Risky, can be impersonated
Other services or protocols	Risky, cannot be impersonated

remove. In contrast, our experiment is performed in a well-controlled experiment environment that makes it much easier to sanitize.

3 Capturing Malware Network Behavior

Contemporary malware relies more and more on the network to achieve its ultimate purpose [9,22]. Malware often downloads binaries needed for its functionality from the Internet or connects into command and control channel to receive instructions on its next activity [10]. Advanced persistent threats [23] and keyloggers collect sensitive information on users' computers but need network access to transfer it to the attacker. DDoS attack tools, scanners, spam, and phishing malware require network access to send malicious traffic to their targets.

We study malware's network activities because they have become essential for modern malware. The first step in this study includes capturing malware's network traffic in an environment, which is transparent to malware, but also minimizes risk to Internet hosts from adversarial malware actions.

3.1 Analysis Environment

We leverage the experimentation platform, called Fantasm [6]. Fantasm is built on the DeterLab testbed [14], which is a public testbed for cyber-security research and education. DeterLab allows users to request several physical nodes, connect them into custom network topologies, and install custom OS and applications on them. Users are granted root access to the machines in their experiments. Fantasm runs on Deterlab with full Internet access, and carefully constrains this access to achieve productive malware analysis, and minimize risk to outside hosts. In our analysis, we run malware on a bare-metal Windows host, without any virtualization or debugger. We capture and analyze all network traffic between this machine and the outside using a separate Linux host, sitting in between the Windows host and the Internet. Both hosts are controlled by the Fantasm framework.

Fantasm makes decisions on which communications to *impersonate*, i.e., intercept and answer itself, which to *forward* and which to *drop*. This decision is made by taking into account each outgoing flow separately, making an initial decision, and revising it later if subsequent observations require this. Fantasm defines a flow as a unique combination of destination IP address, destination port, and protocol. Each flow is initially regarded as *non-essential*, and it is dropped. If this leads to the abortion of malware activity, Fantasm stops analysis, restarts it, and regards that specific flow as *essential*. Fantasm then considers if it can fake replies to this outgoing connection in a way that would be indistinguishable from the actual replies, should the flow be allowed into the Internet. If Fantasm has an *impersonator* for the given destination and the given service, it will intercept the communication and fake the response. Otherwise, it will evaluate if the outgoing flow is risky, i.e., potentially harmful to other Internet hosts. If so, the flow will be dropped. Otherwise, it will be let out into the Internet. Table 1 illustrates the criteria used by Fantasm to determine if a flow is risky or not and if it can be impersonated. Note that even flows considered not risky and let out are still subjected to further monitoring and may be aborted if they misbehave because they could become part of scanning or DDoS. To minimize the risk of unwitting participation in attacks, Fantasm actively monitors for these activities and enforces limits on the number of *suspicious* flows that a sample can initiate. A suspicious flow receives no replies from the Internet. Many scans and DDoS flows will be classified as suspicious. If the analyzed malware sample exceeds its allowance of suspicious flows (10 in the current implementation), Fantasm aborts its analysis.

4 Sample Embedding

Once each sample is analyzed in Fantasm, we extract flow-level details of the malware's communication with the Internet from the captured traffic traces and create an *embedding* for each flow and each sample. Our selected *flow features* can be categorized into three broad categories:

- **Header information.** This information includes destination address, port, and transport protocol. We use this information to detect when different malware samples contact the same server, or same destination port (and thus may leverage the same service at the destination server).
- **Flow dynamics.** This includes a sequence of application-data-units (ADUs, see Sect. 4.1) exchanged in each direction of the flow, which corresponds to request and response sizes on the flow. We use this flow dynamics to detect malware flows with similar communication patterns.
- **Payload information.** We use a frequency-based technique (see Sect. 4.2) to generate an embedding of the flow's content, which can be used for a fast comparison between flows.

To create an embedding for an entire sample we use all its flow embeddings. A detailed list of features selected for each category is provided in Table 2. We next introduce two metrics we will use to closely compare malware samples.

Table 2. Features selected for flow and sample embedding

Feature Category	Feature Selected	Data Type
Header information	Source/Destination address	string
	Source/Destination port	number
	Protocol	string
Flow dynamics	Application data units	list
Payload information	Byte frequency table	dict

4.1 Application Data Unit (ADU)

Flow dynamics include sequences of application data units with their sizes and direction. An *application data unit*, or *ADU*, is an aggregation of a flow by the direction that combines all adjacent packets transmitting in the same direction together, while maintains boundary of direction shift. Intuitively, ADU dynamics seeks to encode the length of requests and responses in a connection between a malware sample and a remote host. A transformation of packet trace to ADU is illustrated in Table 3. The ADU sequence is useful to detect similar flows across different malware samples based on their communication dynamics. For example, two different samples may download the same file from two different servers, and the contents may be encrypted with two different keys. However, the ADU sequence of these two flows should be very similar, enabling us to detect that these two flows have a similar or same purpose.

4.2 Payload Byte Frequency

Payload usually stores application-level data. Not all malware flows carry a payload, but if it is present it usually carries high-level logic, such as new instructions or binaries that are important for new functionality in malware. Hence it is important to study payload contents. On the other hand, payload information usually does not have a specific structure, as different malware may organize its data differently. We thus need a way to quickly summarize and compare payloads that may have very different formats.

We transform each flow's payload into a dictionary that encodes each byte's frequency. Keys to this dictionary are all possible byte values, 0–255, and the values being stored are the counts of how many times the given byte value was present in the payload. Finally, we divide each count with the total payload size to arrive at the frequency of byte values. This encoding has two advantages: First, it has a fixed and much smaller size than the actual payload. Second, it simplifies our similarity comparison between flows and samples.

4.3 Malware Similarity

Another useful application of behavior signatures is to study overlap in behaviors between different malware samples, e.g., to detect polymorphic or metamorphic

Table 3. ADU transformation from packet sequence

Pkt ID	Direction	Pkt size
1	incoming	50
2	incoming	60
3	outgoing	50
4	incoming	100
5	outgoing	70
6	outgoing	90
7	incoming	80
8	incoming	100

(a) Packet sequence

Ref. ID	Direction	Pkt size
(1+2)	incoming	110
(3)	outgoing	50
(4)	incoming	100
(5+6)	outgoing	160
(7+8)	incoming	180

(b) ADU sequence

malware, and to understand how malware functionality evolves, and how common it is across samples.

Obviously, our behavior signatures can yield a wide range of similarity measures, depending on how we define what "similar" means, and what weight we assign to features during the comparison.

Our calculation of sample similarity depends on two distance measures:

1. **Inverse Jaccard Score for ADU comparison**: For a pair of flows, their ADU sequences are compared by calculating the Jaccard score, taking ADU sizes as the values in the set. The Jaccard score between two ADU sequences X and Y is defined in Eq. 1.

$$J(X,Y) = \frac{|X \cap Y|}{|X \cup Y|} \tag{1}$$

The flow similarity based on the original Jaccard score ranges from 0 to 1, with higher values denoting higher similarity. We convert this score into a *distance-type* metric "inverse Jaccard score" – $\overline{JS} = 1 - JS$, with lower values denoting lower distance and thus higher similarity.

2. **Kullback-Leibler divergence for payload comparison**: We compare payloads of flows P and Q by calculating Kullback-Leibler divergence (KL measure) between their payload frequencies, as described in Eq. 2

$$D_{KL}(P||Q) = \sum_{b \in BY} P(b) log(\frac{P(b)}{Q(b)}) \tag{2}$$

where BY is set of all possible byte values and P(b), Q(b) are frequencies of value b in payloads of the flows P and Q respectively. KL measure ranges from 0 to ∞, with lower values denoting higher similarity, i.e. it is already a distance-type metric.

When comparing two samples for similarity, we calculate their distance from flow distance measures in the following way. For each flow in sample A, we compare it with each flow in sample B, and take the lowest distance measure, i.e.

$$d(f_A) = d(f_A, B) = \min_{f_B} d(f_A, f_B) \tag{3}$$

where d is either $\overline{JS}$ or KL measure. We repeat this process for each flow in sample B. Finally, we calculate the **average distance**, which is the average over all flows, i.e.

$$d_{avg}(A, B) = \underset{f_X \in \{f_A, f_B\}}{avg} (d(f_X)) \tag{4}$$

5 Evaluation of Contemporary Malware

We now use features and patterns identified or defined in the previous sections to study malware sample similarity and prevalence of some selected high-level behaviors in contemporary malware.

5.1 Experiment Setup

We build our experiment environment using the Fantasm platform [6] as introduced in Sect. 3.1. Fantasm utilizes the Deterlab infrastructure to construct a LAN environment with a node running Windows to host the malware, and another node running Ubuntu Linux acting as the gateway. In addition, Fantasm provides necessary services for impersonators, and monitors the network activities by capturing all network packets using *tcpdump*. One round of analysis for a given malware sample consists of the following steps:

- Enable service network monitoring on Linux gateway
- Reload operating system on Windows node and set up necessary network configurations
- Deploy and start the malware binary and continue running it for a given period (we used 5 min)
- Kill the malware process and save the captured network trace.

This setting has the advantage that it is immune to current analysis evasion attempts by malware, because it does not use a debugger or a virtual machine. By reloading OS for each run, it ensures that each sample is analyzed in an environment free from any artifacts from the previous analysis rounds.

The malware samples we used in the study were provided by the Georgia Tech Apiary project [1]. We selected malware samples captured throughout 2018 for this research. Next, we submitted each sample to VirusTotal [25] to determine the type of malware, and ensure that the sample is recognized as malicious, from which we randomly picked 8,217 samples. We then analyzed each selected sample in our experiment environment, using the method described above. After the analysis, we have saved traces with all captured communications to and from the Windows node.

6 Clustering Flow Embeddings Using Machine Learning

To evaluate the effectiveness of using flow embeddings and payload byte frequencies on studying malware behavior, we cluster flows based on their metrics and see how well they perform on differentiating traffic behavior. We choose to use OPTICS algorithm (Ordering Points To Identify the Clustering Structure) from Scikit-Learn, which is an algorithm for finding density-based clusters in spatial data. The parameters for OPTICS we use are: $\epsilon = 2$, and min_samples $= 100$.

We cluster malware flows based on two features from our embedding: *10-ADU sequence*, which is useful to identify common communication patterns, and *total payload size, payload size of printable characters, and payload size of all other characters.* We choose to use these higher-level features instead of our payload frequency maps because they are not sensitive to small payload changes.

In total, we have analyzed 200,236 flows from the 6,595 malware samples out of the 8,172 samples we have selected. The remaining 1,577 samples do not successfully exchange payload with an external host. They send only DNS queries to the local resolver but do not initiate any further contact with the outside. In most cases, this happens because malware appears dormant during our analysis. We exclude these samples from further analysis.

6.1 Clustering by ADU Sequence

The clustering based on ADU results in 53 clusters and one group of unclustered flows, containing 8.71% of the flows. We exclude unclustered flows from further analysis. Based on the behavior, the flow clusters can be further grouped into one of the following larger categories:

- HTTP Downloading – incoming traffic volume is larger than outgoing and the application protocol is HTTP.
- HTTP Uploading – outgoing traffic volume is larger than incoming and the application protocol is HTTP.
- UDP Uploading – outgoing traffic volume is larger than incoming and the transport protocol is UDP.
- ICMP Scanning – various external hosts are contacted using ICMP.
- HTTPS Downloading – incoming traffic volume is larger than outgoing and the application protocol is HTTPS.
- HTTPS Uploading – outgoing traffic volume is larger than incoming and the application protocol is HTTPS.
- Unestablished - the flow attempted to connect to an external host but there was no reply or did not finish the 3-way handshake.

We summarize high-level findings based on these categories in Table 4. In this table, we gather information including several clusters that belong to a behavior category, the ratio of clustered flows for the behavior category, the ratio of samples for the behavior category that its flows belonging to, as well as the average $\overline{JS}$ distance for all flows within this behavior category.

Table 4. High-level summary of clustering result using ADU.

Behavior	Cluster Count	Effective Flow ratio	Sample Span Span Ratio	Avg flow $\overline{JS}$ distance (lower is better)	Avg sample $\overline{JS}$ distance (lower is better)
HTTP Downloading	17	2.86%	16.46%	0.1542	0.1717
HTTP Uploading	2	0.21%	1.91%	0.1571	0.2017
UDP Uploading	4	1.17%	1.87%	0.0001	0.1901
ICMP scanning	5	65.02%	2.04%	0.0	0.2120
HTTPS Downloading	7	4.00%	2.07%	0.0749	0.1379
HTTPS Uploading	8	2.90%	2.19%	0.1621	0.1931
Unestablished	9	23.92%	67.57%	0.0034	0.3844

The largest category is ICMP scanning, containing 5 clusters, 65.02% of total flows, and 2.04% of samples. The second-largest category is unestablished flows, containing 9 clusters, 23.92% of flows, and 67.57% of samples[1]. This suggests that most malware samples have many unsuccessful connections. These two types of flows cover 14 clusters, 88.94% of all flows, and span 69.64% of the samples. HTTPS downloading category consists of 7 clusters, 4% of the flows, and 2.07% of the samples; meanwhile, HTTPS uploading category contains 8 clusters, 2.90% of the flows, and 2.19% of the samples. HTTP downloading category contains 17 clusters, 2.86% of the flows, and 16.46% of the samples, while HTTP uploading contains 5 clusters, 0.21% of the flows, and 1.91% of samples. This suggests that more malware samples utilize unencrypted HTTP traffic compared to encrypted HTTPS traffic, which allows for their payload analysis and possibly easier detection and filtering. The rarest observed behavior is UDP uploading, relating to 2 clusters, 0.53% of the flows, and 0.24% of the samples. As Table 4 shows, the average $\overline{JS}$ distance of flows within categories is very low, which means that flows have similar dynamics (similar sizes of the first 10 ADUs). The ADU sequences of clusters could thus be used as behavioral signatures for malware detection.

6.2 Clustering by Payload Sizes

We next cluster flows by payload sizes, resulting in 37 clusters and one group of unclustered flows containing 14.16% of flows. Note that while we cluster by total, printable and non-printable character length of the payload, the average KL distance for flows in each category is very low (ranging from 0.0 to 0.16). This means that flows in the same category have very similar payloads, not just in length but also with regard to the actual byte values.

We again categorize the clusters using the same criteria as we used for ADU clusters. The categories and their coverage of clusters, flows and samples, as well as the average KL distance between flows in each category are shown in Table 5.

[1] A sample can have multiple flows and thus can appear in multiple categories.

We see a similar trending as in ADU sequence-based clusters. ICMP is the dominating type of flows (69.29%) and is only shown in 0.12% of the samples with 0.0 average KL distance because there is no variation in the payload. This is followed by unestablished that contains 25.86% of the flows from 68.81% of the samples with average KL distance of 0.0034, which reinforces the results from ADU analysis that most samples contain failed connection attempts. We also find that more HTTP traffic is detected than HTTPS traffic for both downloading and uploading flows. Payload size also detects more HTTPS uploading flows from 1.92% of all effective flows from 5.26% of the samples, which covers more samples than its ADU equivalence. UDP is still the rarest type of flows detected here with 0.42% of the effective flows from 1.11% of the samples. As mentioned above, the average KL distance is calculated based on the payload frequency map and the result distances are still very low suggesting very high similarity within each cluster. This proves the effectiveness of using the payload frequency map to detect payload patterns.

We now take a closer look at the payload of flows in dominant clusters. We skip unestablished and HTTPS encrypted flows as we cannot analyze their payload. We illustrate our findings through three examples.

Table 5. High-level summary of clustering result using payload sizes.

Behavior	Cluster Count	Effective Flow Ratio	Sample Span Ratio	Avg flow KL distance (lower is better)	Avg sample KL distance (lower is better)
HTTP Downloading	18	2.05%	9.78%	0.0194	2.557
HTTP Uploading	2	0.28%	2.90%	0.0738	0.1397
UDP Uploading	1	0.42%	1.11%	0.2289	2.480
HTTPS Downloading	1	0.09%	0.94%	0.0161	2.879
HTTPS Uploading	5	1.92%	5.26%	0.1283	0.3195
ICMP Scanning	4	69.29%	0.12%	0.0	0.1732
Unestablished	9	25.86%	68.81%	0.0034	2.149

From ICMP ping packets, we observe three dominant types of payload:

- "Babcdefghijklmnopqrstuvwabcdefghi".
- "\u0000" of 28 bytes.
- "b\u0004\u0000EEEEEEEEEEEEEEEEEEEEEEEEEEEEEEEE" (with first character being either "b", or "o" or "\u0601").

This suggests that 150 malware samples in our dataset that perform ICMP scanning may reuse only three different code segments to generate scan packets. ICMP packet payloads we identified can be used as payload signatures for malware detection.

The unencrypted HTTP traffic also shows several dominant behaviors that are interesting. We show them in the Table 6. One type of traffic tries to query and access the root of `google.com` using "GET / HTTP/1.1". It receives "HTTP/1.1 301 Moved Permanently" response but ignores it and keeps sending the same request repeatedly. Such behavior is unusual for benign code and could be used as a behavioral signature to detect malware samples. Another interesting type of HTTP traffic is a type of GET request that tries to retrieve a potentially malicious payload from a malicious website but uses `google.com` as Referrer in the header. We suspect that this may be a way to circumvent the detection of some network defense systems by masquerading as an innocent redirection from Google query. Since a benign code could exhibit similar behavior we cannot use this pattern for malware detection, but it could be used to identify suspicious flows and send them to a more sophisticated defense for further scrutiny.

Another common HTTP flow behavior attempts to query many web sites using fixed-length and random-looking domain names, such as `qexylup.com` or `vowyzuk.com`, with the endpoints being `/login.php` or `/key.bin`. These patterns could be used as behavioral signatures for malware detection.

Table 6. Notable behaviors of HTTP Downloader

HTTP Behavior	Request Example	Flow ratio over HTTP Downloader
Query google.com	GET / HTTP/1.1 Host: google.com	35.83%
Access potential malicious payload on compromised website	GET /.../ddos.bss HTTP/1.1 Host: migsel.com (Migsel bike parts) Referrer: google.com	14.17%
Access potential malicious websie	GET /key.bin HTTP/1.1 Host: qexylup.com Referrer: google.com	39.76%

For UDP flows, we found a type of payload similar to what is used in Bittorrent tracker searching and can confirm that some flows even use the canonical Bittorrent port 6881 for communication. The flows in this cluster have a low average KL distance of 0.2289, indicating a high payload similarity. Thus their payloads could be used to devise a payload signature for malware detection.

6.3 Sample Diversity

We analyze sample diversity based on our findings from flows. We found that 5,223 samples contain flows from different cluster groups from our previous results, which consists of 63.9% of all samples. We also listed the average sample $\overline{JS}$ distance in Table 4 and average sample KL distance in Table 5. We notice that for samples, both the average $\overline{JS}$ distance and average KL distance are higher than their flow counterparts, because those samples may contain flows of different types. We manually inspected some of those samples, and find some types of flows are more likely to be found in the same sample, such as HTTP GET requests for different URLs, HTTP uploading followed by HTTP downloading for potentially malicious payload, etc. These findings suggest that many samples exhibit multiple high-level behaviors and may be multi-purposed.

7 Discussion and Future Work

Our clustering results show that many flows are very similar and that we can use information about their ADU behavior and payload to devise behavioral and payload signatures for contemporary malware. Since the malware ecosystem changes rapidly, the signatures we devise today will likely be obsolete tomorrow. However, our methodology can be used with contemporary malware samples to identify future clusters of behaviors and payloads and to help defenses keep track of malware evolution.

Our current result is still bounded by time and computing power restrictions. Given more time and better infrastructures, we can foresee several future directions to continue our research.

Increase Analyzed Sample Repository. We would like to examine more malware samples and expand our analysis, to balance out samples in each high-level behavior group. We would also like to perform a longitudinal study of malware evolution over time, to quantify how much dominant behaviors change.

Understand Sample Genealogy. Our results can currently help us identify samples that share similar or identical flows, suggesting that these samples may have a common author or that they may share code. We would like to extend our analysis to map out the evolution of malware samples, e.g., which sample came first, how did the specific behavior (e.g., contacting a C&C channel) change over time, etc.

Malware Detection. Most high-level malware behaviors also occur in benign software, which makes them unreliable for malware detection purposes. However, combinations of these high-level behaviors may be unique to malware and could be useful for detection. For example, a software that scans other hosts and then copies data over is unlikely to be benign, although each of these actions separately could be undertaken by benign software (e.g., probing several servers could look like scanning if servers are unresponsive, and data can be uploaded to a cloud for legitimate reasons) As we extend our malware analysis to more samples, we expect to find more behavior combinations, which can be used for malware detection.

8 Conclusion

In this work, we propose to use malware's network traffic patterns to identify high-level behaviors of malware. We define "application data unit" to study malware traffic behavior and "payload frequency map" to study the payload behavior of malware. We then cluster flows from those malware samples based on these features and then form high-level behavior groups. The results show that each behavior group shows a significant behavior pattern. We confirm that flows from each cluster are very similar based on our distance metrics, suggesting that malware may be created by modifying existing code. We then look deeper into the actual behavior patterns within each cluster and find features that can be used to devise behavioral signatures for malware defense. We further analyze sample-to-flow mapping and confirm that 63.9% of the samples contain flows that belong to different behavior clusters, suggesting that contemporary malware is more likely to be multi-purposed.

References

1. Apiary. http://apiary.gtri.gatech.edu/. Accessed 09 May 2018
2. Branco, R.R., Barbosa, G.N., Neto, P.D.: Scientific but not academical overview of malware anti-debugging, anti-disassembly and anti-vm technologies. Black Hat (2012)
3. Chakkaravarthy, S.S., Sangeetha, D., Vaidehi, V.: A survey on malware analysis and mitigation techniques. Comput. Sci. Rev. **32**, 1–23 (2019)
4. Chen, X., Andersen, J., Mao, Z.M., Bailey, M., Nazario, J.: Towards an understanding of anti-virtualization and anti-debugging behavior in modern malware. In: IEEE International Conference on Dependable Systems and Networks With FTCS and DCC, 2008. DSN 2008, pp. 177–186. IEEE (2008)
5. Christodorescu, M., Jha, S.: Static analysis of executables to detect malicious patterns. Technical report, WISCONSIN UNIV-MADISON DEPT OF COMPUTER SCIENCES (2006)
6. Deng, X., Shi, H., Mirkovic, J.: Understanding malware's network behaviors using fantasm. Proceedings of LASER 2017 Learning from Authoritative Security Experiment Results, p. 1 (2017)
7. Egele, M., Scholte, T., Kirda, E., Kruegel, C.: A survey on automated dynamic malware-analysis techniques and tools. ACM Comput. Surv. (CSUR) **44**(2), 6 (2012)
8. Guarnieri, C., Tanasi, A., Bremer, J., Schloesser, M.: The cuckoo sandbox. https://cuckoosandbox.org/ (2012)
9. Holz, T., Engelberth, M., Freiling, F.: Learning more about the underground economy: a case-study of keyloggers and dropzones. In: European Symposium on Research in Computer Security, pp. 1–18. Springer, Cham (2009)
10. Huang, K.T.D.Y., GrierD, D.W.E.B.C., Holt, T.J., Kruegel, C., McCoy, D., Savage, S., Vigna, G.: Framing dependencies introduced by underground commoditization. In: Workshop on the Economics of Information Security (2015)
11. Kaspersky. Kaspersky lab detects 360,000 new malicious files daily. https://www.kaspersky.com/about/press-releases/2017_kaspersky-lab-detects-360000-new-malicious-files-daily (2017). Accessed 23 Sept 2018
12. Kinder, J., Katzenbeisser, S., Schallhart, C., Veith, H.: Detecting malicious code by model checking. In: International Conference on Detection of Intrusions and Malware, and Vulnerability Assessment, pp. 174–187. Springer, Cham (2005)
13. Lever, C., Kotzias, P., Balzarotti, D., Caballero, J., Antonakakis, M.: A lustrum of malware network communication: evolution and insights. In: 2017 IEEE Symposium on Security and Privacy (SP), pp. 788–804. IEEE (2017)

14. Mirkovic, J., Benzel, T.: Deterlab testbed for cybersecurity research and education. J. Comput. Sci. Colleges **28**(4), 163–163 (2013)
15. Morales, J.A., Al-Bataineh, A., Xu, S., Sandhu, R.: Analyzing and exploiting network behaviors of malware. In: International Conference on Security and Privacy in Communication Systems, pp. 20–34. Springer, Heidelberg (2010)
16. Nari, S., Ghorbani, A.A.: Automated malware classification based on network behavior. In: 2013 International Conference on Computing, Networking and Communications (ICNC), pp. 642–647. IEEE (2013)
17. Raghuraman, C., Suresh, S., Shivshankar, S., Chapaneri, R.: Static and dynamic malware analysis using machine learning. In: First International Conference on Sustainable Technologies for Computational Intelligence, pp. 793–806. Springer, New York (2020)
18. Rossow, C., Dietrich, C.J., Bos, H., Cavallaro, L., Van Steen, M., Freiling, F.C., Pohlmann, N.: Sandnet: network traffic analysis of malicious software. In: Proceedings of the First Workshop on Building Analysis Datasets and Gathering Experience Returns for Security, pp. 78–88. ACM (2011)
19. Sharma, A., Sahay, S.K.: Evolution and detection of polymorphic and metamorphic malwares: a survey. *arXiv preprint*http://arxiv.org/abs/1406.7061arXiv:1406.7061 (2014)
20. Shi, H., Alwabel, A., Mirkovic, J.: Cardinal pill testing of system virtual machines. In: USENIX Security Symposium, pp. 271–285 (2014)
21. Shi, H., Mirkovic, J.: Hiding debuggers from malware with apate. In: Proceedings of the Symposium on Applied Computing, pp. 1703–1710. ACM (2017)
22. Stone-Gross, B., Holz, T., Stringhini, G., Vigna, G.: The underground economy of spam: a botmaster's perspective of coordinating large-scale spam campaigns. LEET **11**, 4–4 (2011)
23. Tankard, C.: Advanced persistent threats and how to monitor and deter them. Network Secur. **2011**(8), 16–19 (2011)
24. Ucci, D., Aniello, L., Baldoni, R.: Survey of machine learning techniques for malware analysis. Comput. Secur. **81**, 123–147 (2019)
25. VirusTotal. Virustotal-free online virus, malware and url scanner. https://www.virustotal.com/en (2012)
26. Willems, C., Holz, T., Freiling, F.: Toward automated dynamic malware analysis using cwsandbox. IEEE Security & Privacy **5**(2) (2007)

RC4D: A New Development of RC4 Encryption Algorithm

Rawan Alsharida[1], Maytham Hammood[1(✉)], Mohamed A. Ahmed[1], Barzan Thamer[1], and Mohanaad Shakir[2]

[1] Department of Computer Science, College of Computer Science and Mathematics, University of Tikrit, Tikrit, Iraq
{rawan-adel,maythamhammood}@tu.edu.iq
mohamed.aktham3@gmail.com, sdd3512@gmail.com

[2] College of Business, University of Buraimi, Al-Buraimi, Oman
mohanaad.t@uob.edu.om

Abstract. Cryptography is one of the essential methods for securing the information. In cryptography, there are many encryption algorithms; some of them strong where the others are broken. RC4 stream cipher one of the most common algorithms that are characterized by its speed in implementation does not need large storage space and has less complexity, but there are weaknesses in its output. Numerous researches work on the RC4 stream cipher to boost the security of it, to be strong enough. However, the biases in the output are still in most of the enhancement. The researchers claim that its swap function is responsible for those biases. They recommended to ignore some initial bytes from the keystream output, to dispose of this before de facto encryption begins. This paper present new development over the RC4 algorithm (RC4D) via amendment in the first and second parts of the algorithm. In the first part, it increases the use of the key operations to obtain more considerable randomness, while adding one more random variable and use the Xor function in the second part. Thus, the experimental result of the NIST statistical tests and the distant-equalities statistical analysis shows the RC4D more robust than the original RC4.

Keywords: RC4 · Stream cipher · Cryptography

1 Introduction

Encryption is vital nowadays for keeping our information confidential. We need to maintain our information through it convert into unreadable in the public channel via encryption [1]. Encryption aims to protect information from any attempt to reveal confidential data by any unauthorized third party, and the data is protected through the use of one of the algorithms for encryption. Besides, encryption in information security means encrypting data and making it incomprehensible for protection. Data can only be processed with an encryption key

B. Ghita and S. Shiaeles (Eds.): INC 2020, LNNS 180, pp. 19–30, 2021.
https://doi.org/10.1007/978-3-030-64758-2_2

that converts the regular written form into an encrypted format [2]. There are two primary types of encryption algorithms which are symmetrical and asymmetric. In the asymmetric key type use two key one of them for encryption and the other for decryption. Where in the symmetric key just use one key for encryption and decryption and key should be random. The randomness is significant, and it is challenging to be expected. Here, the strong point of this type, which is meant by randomness according to the definition, and there are two types of randomness; True Random Number Generators(TRNGs) and Pseudo Random Number Generators (PRNGs) [3]. However, the type (TRNGs) that generates randomness using the unpredictable physical process noise likes thermal noise or jitter. This type is not suitable for encryption because the source of randomness is non-deterministic for that reason the same chain cannot be regenerated but it is good to generate a random keys for encryption. Where the second type (PRNGs) generates a certain sequence of numbers by using a specific function and mathematical algorithms to generate a pseudo-random series in a deterministic way that will appear random, however, this type is used in encryption because it is fast, and can regenerate the same sequence by using the same input string as a seed. The seed must be random and long enough to increase the randomness of the output sequence. RC4 is an algorithm consisting of a simple structure that has been proven after being subjected to necessary testing and analysis by researchers and has proven to be powerful enough on various platforms. The fundamental of this algorithm is one internal state table of size N (which mostly 256), which ultimately functions as the S-box for performing the primary mixing in bytes.

The RC4 algorithm was initially designed in 1987 by Ronald Rivest and was kept as a secret until it was disclosed onto the network in 1994 [4]. This Stream Cipher algorithm, which uses the symmetric key, is an algorithm that is known by its speed, and when comparing it with the DES algorithm, it is the best in terms of speed and more straightforward in terms of complexity. In addition, it is used in various protocols, and multiple applications such as Wi-Fi protected access (WPs) and Transport Layer Security (TLS) for web browsing, emailing, and instant messaging. Many encryption operations used in wireless rely on symmetric encryption and are used in Oracle, SQL, protocol (SSL), Wired equivalency privacy protocol. In this research, the researchers design an encryption algorithm that is characterized by its speed in implementation does not need large storage space, less complicated, and is more reliable in protecting information or data. It also has a higher degree of security for devices with limited resources, and there is no bias for the first bits as in the RC4 algorithm. The remainder of the paper is ordered as comes: Literature Review in Sect. 2. Section 3 is RC4 explanation where Sect. 4 presents some weaknesses of RC4. The presentation of the modified RC4 algorithm is in Sect. 5. Performance evaluation is in Sect. 6. The implementation in Sect. 7 then ended by the conclusion in Sect. 8.

2 Literature Review

Many researchers in the field of data security found weaknesses in the security of the RC4, while many researchers tried to develop the RC4 algorithm to bypass these vulnerabilities. However, an attack on this algorithm was provided by Flunrer et al. [5], and the main goal is to eliminate the interconnection between outputs.

Thus, the creation of several modified algorithms was developed for the original RC4 algorithm and on that disability, but these algorithms that improve the original algorithm affected the speed of implementation. On the other hand, others tried to improve the speed of the algorithm but caused a reduction in randomness [6–18]. Nishith et al. developed the RC4 algorithm order to enhance the security of the RC4 by replacing along these variable lines above it to encode an element. The time it takes to encode and decode using the proposed algorithm is barely a little more substantial than the original algorithm [7]. Hammood et al. [8] introduced a random RC4 algorithm (RRC4) to pass the overcome the weak scheduling of keys in the RC4 algorithm. This algorithm enhanced the confidentiality of the encrypted text and proved more random than the original RC4 algorithm. Also, the coding and decoding time for the improved algorithm was the same time in the original algorithm. Ali et al. [9] presented an advanced algorithm based on the original RC4 algorithm, where this algorithm led to an increase in the randomness generated by using mathematical operations, which are Factorial and some other variables. In this case, the security of the algorithm was enhanced, but in contrast, during the comparison of the improved algorithm with the original algorithm, the results showed there is a clear increase in implementation time and the storage space, where the reason for this increase is the Factorial process. Variably Modified Permutation Composition (VMPC) was suggested by Zoltak's [10], which includes many changes in the initialization and the processing rounds as well as in the output generating. It was contagious to be effective in the application and solution of vulnerability in the KSA of RC4 which was described by Fluhrer et al. [5] Pseudo-Random Generation Algorithm (PRGA) structure of VMPC is sophisticated more than the RC4 which makes it higher resistant to attack. Paul and Preneel [11] through their research, introduced a new algorithm, an improved RC4A algorithm, and as a reinforcement over the RC4 algorithm, after discovering an unusual statistical weakness in generations of the first two key bytes for RC4. They said that the number of outputs demanded to distinguish an RC4 random output was 225. RC4A is powerful against most of the weaknesses in RC4, especially the weaknesses of distribution at the first two bytes output. Nevertheless, one year later Maximov introduced a distinct attack on VMPC and RC4A together. Thus distinct attack could distinguish ciphertext from a truly random number [12]. Pu and Chung, [13] presented a group key update method to improve the randomness of RC4 thus increase the security against the attacks by using the same length of the key and initialization vector. The experimental results show this method uses minimum resource from constrained devices such as wireless sensors. The results confirm that the new produced key and the update of the key are random. Also, the randomness will improve during the abundance of sensor nodes is increased.

Hammood et al. presented [14] new improvement over RC4 by using two-state tables to rise its randomness, thus reduce the correlation between the output and the key but also the time of encryption is increased. Yao et al. [15], through his research, contributed analysis and improvement of the safety of the RC4 algorithm via applying a public key with RC4. Still, it drove to an increment in the system size and execution time. Three algorithms presented in [16] for improving RC4 each one has different properties but RC4-2S+ with two state tables to generate four key bytes in every round are the best one and more randomness without affecting its speed, but it has high bias. Suman et al [17] analysed different variants of RC4 to find out the benefit of dropping initial bytes. Also, they produced multiple S-boxes by various logics and a unique key to increasing robust of RC4 but the speed of RC4 increased. Zelenoritskaya et al. [18] presented an improvement of the RC4, by using parallel LFSR with stochastic transformation boxes. Thus, the complexity of the relationship between the key and the initial state increased. However, the using of LFSR led to an increase in the time of execution.

3 Description of RC4

Several Stream Cipher algorithms rely on the use of Linear Feedback Shift Registers (LFSRs), particularly in physical components, but when the RC4 algorithm was de-signed (LFSRs) were not used. RC4 algorithm involves two parts to produce the first component key. It is a Key Scheduling Algorithm (KSA), as shown in Algorithm 1. The second component is the Pseudo-Random Generation Algorithm (PRGA), as shown in Algorithm 2, which is executed consecutively [4].

RC4 algorithm has a variable key size domain between 8 to 256 bytes to initialize 256 bytes in the initial state array by the number from 0 to 255. The KSA algorithm produces flipping or initial switching by mixing the flipping or the corresponding switch by using the key, and the result of this flipping is considered an entrance to the part PRGA. This step creates the final key RC4 starts with switching, and the secret key is used to produce random swapping. The next phase is the PRGA algorithm that generates bytes of the keystream, which performs the XOR process with the plaintext to obtain the ciphertext. The concept of the RC4 algorithm is to make the flipping or switching of elements by exchanging them with each other for higher randomization.

Algorithm 1. RC4 - KSA algorithm

```
Set N ← 256
Set i ← 0
while (i<N) do
    S[i] ← i
    i ← i + 1
end while
Set i ← 0
Set j ← 0
while (i<N) do
    j ← (j + S[i] + K[i mod L])mod N
    Swap (S[i] , S[j])
end while
return S
```

Algorithm 2. RC4 - PRGA algorithm

```
Set i ← 0
Set j ← 0
while generate key-stream do
    i ← (i + 1)mod N
    j ← (j + S[i])mod N
    Swap (S[i] , S[j])
    Output ← S[(S[i] + S[j])mod N]
    Ciphertext ← Output ⊕ Plaintext
    return Ciphertext
end while
```

4 The Cryptanalysis of RC4

Cryptanalysis is used to violate cryptographic security systems by breaking encrypted messages without the known cryptographic key. It mainly focuses on identifying the non-random procedure. There are many weaknesses in the RC4 algorithm a set of vulnerabilities that can be solved, and the other group is considered a threat by attackers to exploit. One of the weaknesses in the case of initialization is the statistical bias that occurs in word distribution in the first outlet. The problem of the algorithm in the first component (KSA) is simple but random. There is a relationship between some byte units of the secret key, so the analysis of these bytes makes it possible to attack RC4. The key flow that starts at the beginning of the algorithm is a one-time swaps S (identical to the value of (i) the pointer indicating the entry) for low values, and S [j] = j is possible during initialization [19]. Roos [20] found another weakness in RC4, which is a correlation among the values of the first table and the keystream. The primary reason is the status table that started in the sequence (0, 1, 2, ...,

255) and at least one of the 256 temporary keys. The first byte that was created from the key is highly correlated with a few first bytes of PRGA output; thus, to solve this issue, it has been advised to discard the initial byte of PRGA output. The main objective of the adversary is to obtain the original key, internal state, or output key to access the plaintext. Since prior studies of RC4, there are some weaknesses in it, such as biased bytes, and key recovery where various attacks focus on obtaining the private key for the internal state [21]. Tomašević et al. [22] displayed a cryptanalytic attack that utilizes the tree design of this cipher and offers an abstraction of prevailing conditions for the information of its state table. To obtain the initial state will using the hill-climbing strategy for the search in the tree of general conditions. The complication of this attack is weaker than that of an exhaustive search. The successful attack on RC4 utilized the existence of biases in the RC4 keystream to make plaintext recovery attacks against TLS-RC4 [23]. Hammood et al. [24] evaluated the likelihood of the generating ciphertext for each pair of byte values for each 256-byte round than found new four biases.

5 Modified RC4 Algorithm

Due to the existence of many weaknesses in the RC4 algorithm, as mentioned earlier, in this research, the RC4 algorithm was improved through modification in the first and second parts. In the first part, as shown in Algorithm 3, RC4D-KSA was modi-fied through an increase using the key and generates new random variable to obtain more randomness, while in the second part, as shown in Algorithm 4; RC4D -PRGA the algorithm was modified by generate new random variable and XORed it with the random j then find $S[S[j] \oplus S[T]]$. Thus, the randomness will increase. The modifi-cation in the RC4D- KSA does not take much time, because they only happen 265 times.

Algorithm 3. RC4D - KSA algorithm

Set $N \leftarrow 256$
Set $i \leftarrow 0$
while (i<N) **do**
  $S[i] \leftarrow i$
  $i \leftarrow i+1$
end while
Set $i \leftarrow 0$
Set $j \leftarrow 0$
while (i<N) **do**
  $j \leftarrow (j + S[(i + K[i \ modL])modN] + K[i \ modL]) \ modN$
  Swap $(S[i]\ ,\ S[j])$
end while
return S

Algorithm 4. RC4D - PRGA algorithm

Set $i \leftarrow 0$
Set $j \leftarrow 0$
Set $D \leftarrow 0$
while *generate key-stream* **do**
  $i \leftarrow (i+1) mod\ N$
  $j \leftarrow (j+S[i]) mod\ N$
  Swap $(S[i]\ ,\ S[j])$
  $Output \leftarrow S[(S[i]+S[S[j] \oplus S[D]]) mod\ N]$
  Set $D \leftarrow Output$
  $Ciphertext \leftarrow Output \oplus Plaintext$
  return *Ciphertext*
end while

6 Performance Evaluation

There is a set of different criteria for measuring the level of safety and performance of the specific encryption algorithm. At this time, two measurements were used. The first test was distant-equalities statistical test [25] and the second test a random statistical Suite test [27] created by the NIST (National Institute of Standards and Technology). In addition, evaluated the time of the performance of the proposed algorithm.

6.1 Distant-Equalities Statistical Test

As mentioned, there are weaknesses in Stream Cipher algorithms in the process of generating keys; there is bias in randomization. This test works to measure this bias by generating a large number of keys (N) and then measures how similar these keys are between them [25,26]. Simply this test each key with the eight keys followed it's by a matrix consisting of 8 different events: Event Z_i for $i \in 1,2,3,4,5,6,7,8$ The first event Z_1 test the first key with the second key, Event two Z_2 test the first key with the third key, Event three Z_3 test the first key with the fourth key, Event four Z_4 test the first key with the fifth key, Event five Z_5 test the first key with the sixth key, Event six Z_6 test the first key with the seventh key, Event seven Z_7 test the first key with the eighth key, Event eight Z_8 test the first key with the ninth key. Note t : (start test) begins from K1 to Kn. Z_i++ in every t for i =1,2,3,4,5,6,7,8. The probability of each event in the output is p = 1/N, where N is the keyword size of the algorithm. However, the expected number of occurrences for every event of n samples is present in Eq. 1. [25]

$$E = np \tag{1}$$

In addition, the standard deviation of the number of occurrences is given by the binomial distribution as shown in Eq. 2 [25].

Table 1. Distant-equalities statistical test results for RC4.

Samples	Event 1	Event 2	Event 3	Event 4	Event 5	Event 6	Event 7	Event 8
2^{25}	25.9838	−65.9029	26.6885	−24.9691	13.7112	0.489631	4.57946	2.93726

$$\sigma = \sqrt{(np(1-p))} \tag{2}$$

The test calculations the numbers of occurrences for the Events from one to eight then compares them with the results of model E = np utilize σ as the measure of deviation as shown in Eq. 3. [25]

$$d = (f - E)/\sigma \tag{3}$$

Where d is the test calculates, and f represents the measured number, which is standard statistical practice. Table 1 represents the amounts of standard deviations that measured numbers of appearing for Events from Event1 to Event8 were different compere with their estimated values (d = (f − E)/σ) for different word sizes (N = 8) operates. We found a clear bias between the Event1 and the Event2 that the Event2 approximates 3σ of the Event1 almost three times the first event while otherwise, the bias may not be in the Event6 and Event7.

Table 2. Distant-equalities statistical test results for RC4D.

Samples	Event 1	Event 2	Event 3	Event 4	Event 5	Event 6	Event 7	Event 8
2^{25}	1.86613	−2.27851	−1.36554	−0.271437	2.04778	1.33317	0.979784	−0.887913

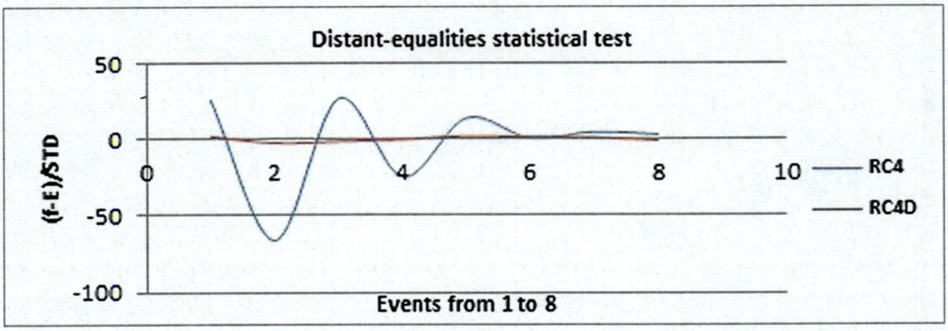

Fig. 1. Distant-equalities statistical test results for RC4 and RC4D.

It was observed that when applied the test over original RC4, the bias is very clear, in most the events spatially in the first five events as shown in Table 1. In contrast, the results of RC4D showed it's very good there is no bias when use the same amount of key as shown in Table 2. The results indicate the randomness increases because the authors add a new random parameter. Thus, the RC4D is given better results than RC4. As shown Fig. 1 which describes the difference between RC4 and the new RC4.

6.2 Randomness Test

The generated keystream is examined by a set of NIST (National Institute of Standards and Technology) tests, which is a statistic set of random number generation tests that include 16 statistical analyses to perform randomization of the output chain of a random number or an integer random number generator [27]. PRGA tests were performed using NIST, and the probability of a valid random number generator was represented with a value of P. These tested are produced and the simplified mean of the P-values from these tests, as shown in Table 3. In the test, the P-values are compared to 0.01, the P-values are passed when they are larger than 0.01, and the series produced is random and uniformly distributed. If the tests give a value of P equal to 1, then the string is a completely random series, but if the P-values is zero, it means that the string is entirely random. Success means that the string is acceptable and has good randomness,

Table 3. NIST Tests Applied to Standard RC4 and Modified RC4 (RC4D).

Statistical Test Name	RC4		RC4D	
	p-value	Result	p-value	Result
Approximate Entropy	0.436169	Pass	0.541461	Pass
Block Frequency	0.325088	Pass	0.438262	Pass
Cumulative Sums (Forward)	0.581981	Pass	0.602363	Pass
Cumulative Sum (Reverse)	0.448262	Pass	0.638262	Pass
FFT	0.457888	Pass	0.497533	Pass
Frequency	0.598393	Pass	0.625319	Pass
Lempel-Ziv compression	1	Pass	1	Pass
Linear Complexity	0.422196	Pass	0.590875	Pass
Longest Runs	0.430753	Pass	0.571836	Pass
Non periodic Templates	0.472041	Pass	0.534455	Pass
Overlapping Template	0.476283	Pass	0.451715	Pass
Random Excursions	0.50275	Pass	0.552624	Pass
Random Excursions Variant	0.588278	Pass	0.506814	Pass
Rank	0.493697	Pass	0.536438	Pass
Runs	0.406088	Pass	0.568629	Pass
Serial	0.558872	Pass	0.450122	Pass
Universal Statistical	0.557777	Pass	0.697207	Pass

while failure indicates that it is not acceptable and not random. Some of the 16 tests accepted large sizes of output series and failed in the small size, as well as other tests accepted large and small sizes of output series. In our program, a large volume (1000000 bytes) of each secret key is created. In this paper, as shown in Table 3 the result indicates that the RC4D is better than the original RC4 in most of the tests.

7 Implementation

When implementing the algorithm using the C++ language, the values in the algorithm inputs in the initial state from 0 to 255, and the key length was 16 byte only. The time of execution in both algorithm the original RC4 and RC4D is approximately the same as shown in Table 4.

Table 4. Execution time of RC4 and RC4D.

Key Size (kilobyte)	RC4 Time (m.s)	RC4D Time (m.s)
100,000	12.488	12.518
1,000,000	117.849	118.315

8 Conclusions

In this paper, the author presents an improvement over the RC4 encryption algorithm. It is also more reliable in protecting information or data and has a higher degree of security for devices with limited resources; also, there is no bias for the first bits as in the original algorithm With the increasing of modern technology nowadays the challenge is how to secure the data and confidential information against the weak souls, which are electronic armies and malicious intentions have piqued these companies. For fear of theft of these data and information has been encrypted, this data and information using different encryption algorithms, but these algorithms contain vulnerabilities exploited by hackers to steal data and information. Thus, to increase security and reduce potential attacks should be used robust encryption algorithms. In this research, we presented improvement over RC4 and test the result by NIST statistical test and a distance statistical test. These tests help to detect the weaknesses of the stream cipher algorithms. The test is distinguished from the previous tests at its speed, as well as it is characterized by giving results of high accuracy and efficiency, so it is useful in the process of improving the cipher algorithms. These characteristics of RC4D algorithm make it a better nominee for effective applications as compared to the original RC4 Algorithm.

References

1. Rueppel, R.A.: Analysis and design of stream ciphers. Springer, New York (2012). https://doi.org/10.1007/978-3-642-82865-2
2. Lamba, C.S.: Design and analysis of stream cipher for network security. In: Second International Conference on Communication Software and Networks, Singapore, pp. 562-567 (2010). https://doi.org/10.1109/ICCSN.2010.113
3. Thomas, D.B., Luk, W.: High quality uniform random number generation using LUT optimised state-transition matrices. J. VLSI Signal Process **47**, 77–92 (2007). https://doi.org/10.1007/s11265-006-0014-9
4. Rivest, R.L.: The RC4 encryption algorithm, RSA Data Security Inc., **129**-2, March 1992. This document has not been made public
5. Fluhrer, S., Mantin I., Shamir, A.: Weaknesses in the key scheduling algorithm of RC4. In: Proceedings of Annual Workshop on Selected Areas in Cryptography, Springer, Heidelberg, vol. 2259, pp. 1-24 (2001). https://doi.org/10.1007/3-540-45537-X_1
6. Xie, J., Pan, X.: An improved RC4 stream cipher. In: International Conference on Computer Application and System Modeling (ICCASM), IEEE vol. 7, pp.156-159 (2010). https://doi.org/10.1109/ICCASM.2010.5620800
7. Sinha, N., Chawda, M., Bhamidipati, K.: Enhancing security of improved RC4 stream cipher by converting into product cipher. Int. J. Comput. Appl. (0975 - 8887) vol. 94 - No .18, pp. 17–21, May 2014. https://doi.org/10.5120/16459-6132
8. Hammood, M.M., Yoshigoe, K., Sagheer, A.M.: RC4 stream cipher with a random initial state. In: Proceedings in Information Technology, Springer, Dordrecht, p. 407 (2013). https://doi.org/10.1007/978-94-007-6996-0_42
9. Sagheer, M.A., Searan, S.M., Alsharida, R.A.: Modification of RC4 algorithm to increase its security by using mathematical operations, J. Software Eng. Intell. Syst. **1**(2), ISSN 2518-8739 (2016). http://www.jseis.org/Volumes/Vol1/V1N2-1.pdf
10. Zoltak, B.: VMPC One way Function and Stream Cipher, Fast Software Encrypt. FSE, LNCS, 3017, pp. 210–225. Springer, New York (2004)
11. Paul, S. Preneel, B.: A new weakness in the RC4 keystream generator and an approach to improve the security of the cipher. In: Fast Software Encrypt, FSE, LNCS 3017. New York: Springer, Heidelberg, pp. 245–259 (2004). https://doi.org/10.1007/978-3-540-25937-4_16
12. Maximov, A.: Two linear distinguishing attacks on VMPC and RC4A and weakness of the RC4 family of stream ciphers. In: Fast Software Encryption, FSE, Vol 3557, pp. 342-358, Springer, Cham (2005). https://doi.org/10.1007/11502760_23
13. Pu, C., Chung, W.C.: Group key update method for improving RC4 stream cipher in wireless sensor networks. In: International Conference on Convergence Information Technology, Gyeongju, pp. 1366-1371 (2007)
14. Hammood, M.M. Yoshigoe, K., Sagheer, A.M.: RC4-2S: RC4 stream cipher with two state tables. In: Information Technology Convergence, pp. 13–20. Springer, Netherlands (2013). https://doi.org/10.1007/978-94-007-6996-0_2
15. Yao, Y., Chong, J., Xingwei, W.: Enhancing RC4 algorithm for WLAN WEP protocol. In: Control and Decision Conference (CCDC), pp. 3623–3627, IEEE (2010). https://doi.org/10.1109/CCDC.2010.5498536
16. Hammood, M.M., Yoshigoe, K., Sagheer, A.M.: Enhancing security and speed of RC4. Int. J. Comput. Network Technol. **3**(02) (2015). https://doi.org/10.12785/ijcnt/030201

17. Das, S., Ghosh, R., Pal, R.K.: An approach of refining RC4 with performance analysis on new variants. Sādhanā **44**(11), 223 (2019). https://doi.org/10.1007/s12046-019-1209-7
18. Zelenoritskaya, A.V., Ivanov, M.A., Salikov, E.A.: Possible modifications of RC4 stream cipher. In: Advanced Technologies in Robotics and Intelligent Systems, pp. 335–341. Springer, Cham (2020). https://doi.org/10.1007/978-3-030-33491-8-40
19. Mister, S. Tavares, S.E.: Cryptanalysis of RC4-like Ciphers. In: International Workshop on Selected Areas in Cryptography, Springer, Cham (1998). https://doi.org/10.1007/3-540-48892-8_11
20. Roos, A.: A class of weak keys in the RC4 stream cipher, September 1995. http://agreg.dnsalias.org/Luminy/WeakKeys-report.pdf
21. Jindal, P., Singh, B.: Performance analysis of modified RC4 encryption algorithm. In: Recent Advances and Innovations in Engineering (ICRAIE), 2014. IEEE (2014). https://doi.org/10.1109/ICRAIE.2014.6909247
22. Tomašević, V., Bojanić, S., Nieto-Taladriz, O.: Finding an internal state of RC4 stream cipher. Inf. Sci. **177**(7), 1715–1727 (2007)
23. AlFardan, N., Bernstein, D.J., Paterson, K.G., Poettering, B., Schuldt, J.C.: On the security of RC4 in TLS. In: Presented as Part of the 22nd USENIX Security Symposium, pp. 305-320 (2013)
24. Hammood, M.M., Yoshigoe, K.: Previously overlooked bias signatures for RC4, 4th Inter-national Symposium on Digital Forensic and Security (ISDFS). Little Rock, AR **2016**, 101–106 (2016). https://doi.org/10.1109/ISDFS.2016.7473526
25. Zoltak, B.: Statistical weaknesses in 20 RC4-like algorithms and (probably) the simplest algorithm free from these weaknesses - VMPC-R, IACR Cryptology ePrint Archive, IJCSN, 2014. https://eprint.iacr.org/2014/315.pdf
26. Zoltak, B.: Statistical weakness in Spritz against VMPC-R: in search for the RC4 replace-ment. IACR Cryptology ePrint Archive, 985, (2014). https://eprint.iacr.org/2014/985.pdf
27. Rukhin, A., Soto, J., Nechvatal, J., Smid, M., Barker, E., Leigh, S., Levenson, M., Van-gel, M., Banks, D., Heckert, A., Dray, J., Vo, S.: A statistical test suite for random and pseudorandom number generators for cryptographic applications, NIST special publication 800–22. Gaithersburg, National institute of standards and technology (NIST) (2001). https://apps.dtic.mil/dtic/tr/fulltext/u2/a393366.pdf

A Novel Multimodal Biometric Authentication System Using Machine Learning and Blockchain

Richard Brown[1], Gueltoum Bendiab[2], Stavros Shiaeles[2](✉), and Bogdan Ghita[1]

[1] CSCAN, University of Plymouth, Plymouth PL4 8AA, UK
richard.brown@students.plymouth.ac.uk , bogdan.ghita@plymouth.ac.uk
[2] Cyber Security Research Group, University of Portsmouth, Portsmouth PO1 2UP, UK
gueltoum.bendiab@port.ac.uk, sshiaeles@ieee.org

Abstract. Secure user authentication has become an important issue in modern society as in many consumer applications, especially financial transactions, it is extremely important to prove the identity of the user. In this context, biometric authentication methods that rely on physical and behavioural characteristics have been proposed as an alternative for convolutional systems that rely on simple passwords, Personal Identification Number or tokens. However, in real-world applications, authentication systems that involve a single biometric faced many issues, especially lack accuracy and noisy data, which boost the research community to create multibiometric systems that involve a variety of biometrics. Those systems provide better performance and higher accuracy compared to other authentication methods. However, most of them are inconvenient and requires complex interactions from the user. Thus, in this paper, we present a multimodal authentication system that relies on machine learning and blockchain, intending to provide a more reliable, transparent, and convenient authentication mechanism. The proposed system combines tow important biometrics: fingerprint and face with age, and gender features. The supervised learning algorithm Decision Tree has been used to combine the results of the biometrics verification process and produce a confidence level related to the user. The initial experimental results show the efficiency and robustness of the proposed systems.

Keywords: Authentication · Machine learning · Blockchain · Multimodal · Security

1 Introduction

Currently, user authentication has become one of the greatest challenges facing the digital world. Traditional authentication methods that rely on tokens, password, and Personal Identification Number (PIN) are gradually becoming obsolete [6]. In fact, tokens and PIN/Passwords offer limited protection and can

B. Ghita and S. Shiaeles (Eds.): INC 2020, LNNS 180, pp. 31–46, 2021.
https://doi.org/10.1007/978-3-030-64758-2_3

be easily lost, stolen, forgotten, guessed, or compromised [7,19]. In this context, the last report by the World Economic Forum [24] revealed that 80% of security breaches, in 2020, are perpetrated from weak and stolen passwords. Moreover, the report affirms that, for companies, 50% of IT help desk costs are allocated to passwords resets, with average annual spend over $1 million for staffing alone [24]. These shortcomings have led to biometric authentication becoming the focus of the research community in last years. It refers to the technology that identifies and authenticate individuals in a fast and secure way through the use of unique behavioural and biological characteristics like fingerprints, hand geometry, vein, face, iris, voice, palm, DNA, etc. [7]. This technology has quickly established itself as an alternative to Personal Identification Number (PIN), tokens and Passwords for various reasons [1]. Biometrics are unique for individuals and almost impossible to replicate or forge [19], which provides superior accuracy and prevent unauthorised access from those who may have the means to steal passwords or PINs [1,7]. Also, Biometric authentication offers convenience, accountability, and reduces the overall administrative costs by eliminating the time consuming to reset passwords [7]. Moreover, they are resistant to social engineering attacks, especially phishing attacks.

Biometric technology has been considered by the research community as the most reliable and safe method for individuals authentication and several biometric systems based on common biological and behavioural characteristics (e.g. fingerprint, face, iris, handwriting, palm, keystroke, etc.) have been developed during last decades [11,19]. As shown in Fig. 1, all biometric systems fellow the same process. First, a biometric system (e.g. fingerprint scanner, digital camera for face, etc.) is used to capture and records a specific trait of the user. The collected biometrics are examined and converted to a template that can be stored in a database, or a smart card [11]. This step is called enrolment. Then, each time the user request access to the system, presented biometric values are compared against these in the stored template. This verification process generates a matching score that designates the degree of similarity between the two biometrics data. The resulting score should be high for legitimate users and low for those from different ones. Based on the obtained matching score (i.e., confidence level), legitimate users are allowed access to the system, while the impostors are rejected. In this step, a biometric sensor is used to extract the trait being used for identification.

In real-world applications, biometric authentication systems which involve one single biometric trait for enrolment and verification are facing a variety of problems such as lack of accuracy due to noisy data, spoof attacks, non-universality, lack of uniqueness, etc. [7]. To address these limitations, many multimodal biometric systems that combine more than one physiological and/or behavioural biometrics have been proposed. [7,11]. Usually, these systems involve a variety of biometrics that are fused, normalised and fed into a machine learning classifier to drive a decision [17]. This led to a highly accurate, secure authentication system. They also provide better performance compared with unimodal

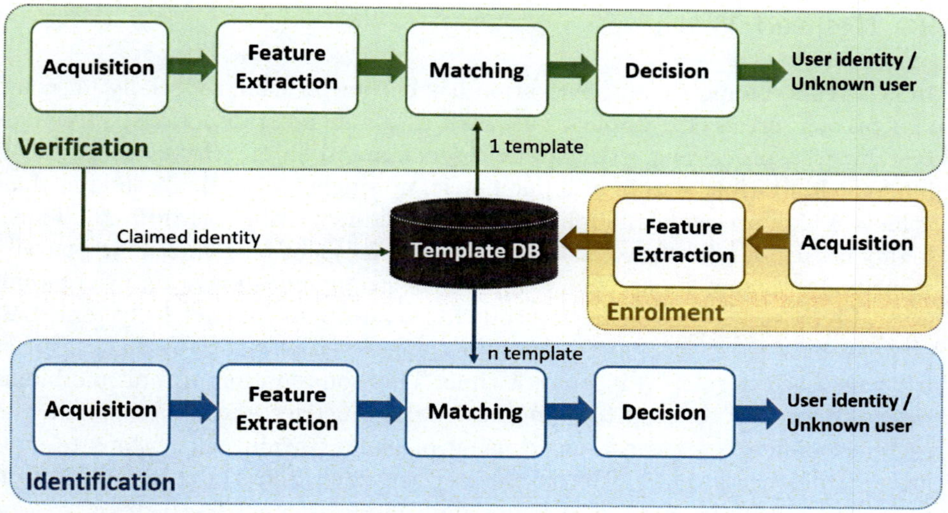

Fig. 1. Biometric authentication process [15]

systems. However, most existing multimodal biometric systems are inconvenient and relied heavily user interaction to authenticate.

In order to fulfil the objectives of a more secure, and transparent authentication mechanism, this paper introduces a novel system for individual identity management that uses multimodal biometric authentication system with machine learning and blockchain. The multimodal biometric system combines four different features for enrolment, identification, and verification. The biometrics are fingerprints, facial. While, age and gender features are driven from the facial biometrics. This will increase the authentication security and overcome the limitations of unimodal systems. Based on the outputs of the biometrics verification process, the supervised learning algorithm decision tree is used to identify a confidence level related to the user.

This confidence level should be high enough to allow the user accessing the protected resources on the web server (i.e. Service Provider). Whereas the blockchain is used to store user details and key data that can be used by the service provider (i.e. the web server) and the identity provider (i.e. BCA server) to encrypt/decrypt the user access token. Further, compared to other multi modal mechanisms, the proposed system minimise the total amount of interaction required for identification and authentication.

The structure of the paper is as follows. Section 2 gives an overview of biometric authentication systems, Sect. 3 describes the general architecture of the proposed methodology used to ensure the user identity during time, while Sect. 4. Finally, Sect. 5 presents the conclusions of this work and proposes future work.

2 Related Work

In last years, biometric authentication has become crucial, especially in security and privacy preserving applications such as, financial transactions, surveillance system, visa processing, critical environments and so on. However, due to the inherent limitations within each biometric, no single biometric method is able to achieve a high precision and reliability of individuals authentication [23]. Thus, in highly critical applications, a single biometric may not be sufficient to guarantee security, but it may be necessary to perform strong authentication by combining several biometrics [22,23]. In this context, several multibiometric systems based on conventional physical and behavioural characteristics such as fingerprint and iris have been developed in present time. This combination of multiple biometrics is commonly referred to as multimodal biometrics authentication. In these systems, biometrics are combined using machine learning algorithms to generate a confidence level, which will be used to either allow or deny access to the requested resources. One of the first multimodal biometrics systems was proposed by Clark NL [3], using a combination of secret knowledge and biometric-based techniques to create an Intelligent Authentication Management System (IAMS). This method used a confidence level, which continuously updated to control the user access to protected resources. This can help in countering the increasing vulnerability of traditional knowledge-based techniques. Our system shares many aspects with this proposal in regard to the confidence level and the use of multiple authentication techniques. With the main goal of creating a system that is robust, secure and does not interfere with the convenience of users.

In a previous work [18], face and speaker recognition modalities are used in a serial mode, where the output of one biometric modality is used to reduce the number of possible individuals that will be checked with the second biometric. Final decision is given by the second biometric from the reduced subset of individuals. This method achieved a low False Rejection Rate (FRR) (3.9%), however, the time consumption is important compared to the fusion method. Thus, most recent multimodal biometric systems have been used the fusion method to combine the features obtained from multiple biometrics. In this context, the fusion has been applied at three different levels: at the features level, at the score matching level or at the decision level. For instance, the multibiometric approach proposed by A. Tharwat et al. [22], has been explored two different fusion methods: fusion at the image level and a multilevel fusion method to combined ear and finger knuckle biometrics. The experimental results showed that the fusion at the image level can improve the overall performance of the authentication system. This method combines the ear and finger knuckle images before extracting the features that will be used by the classification module to produce an abstract value or rank. Authors highlighted that there are several methods for successfully implement a multimodal system, although the paper does not cover how the user is expected to communicate with the system. In addition, having a user take an image of their knuckles and ear is not a user-friendly approach.

In a recent work, J. Peng et al. [18] have been proposed a multibiometric authentication system that combines four finger biometric traits: finger vein, fingerprint, finger shape and finger knuckle print. A score-level fusion method has been used to produce the overall score or confidence level of the target user based on triangular norm. The experimental results showed that the used fusion method obtained a larger distance between honest and imposter score distribution as well as achieves lower error rates. In more recent work [10], T. Joseph et al. have been proposed a multimodal authentication system by fusing the feature points of fingerprint, iris and palm print biometrics. After fusing the features extracted from these biometric modalities, a secret key is generated in two stages and converted into a hash value using MD-5 hashing algorithm. A novel feature-level fusion method has been proposed by Asst. Prof. Masen M et al. to combine face and iris features [16]. First, the face and iris traits are extracted independently using 2D wavelet transform and 2D Gabor filters, respectively. After that, the proposed fusion method is applied by using both canonical correlation and serial concatenation. Then, the deep belief network is used for the verification process. This approach has been validated on the SDUMLAHMT database [16] and achieved an overall recognition accuracy up to 99%. However, the Equal Error Rate (EER) and fusion time are important in comparison with other systems. Many other multibiometric authentication systems have been proposed in last years by using different biometrics and different fusion methods [9,12,20], however, most of them are inconvenient and relied heavily user interaction to authenticate.

3 Proposed Approach

This section presents the detail about the proposed multimodal biometric system for individual's authentication using machine learning and blockchain. As shown in Fig. 2, the authentication process involves three entities: the user, the Service Provider (i.e. web or resource server) and the Identity Provider (i.e. Biometric Confidence Authentication (BCA) server). The BCA server is responsible on the enrolment, identification, and verification of the user biometrics along with his level of confidence. It provides Single Sign-On (SSO) for multiple web applications. Users can monitor their confidence level and submit biometrics through a web interface (i.e. client) provide by the BCA server.

For accessing protected resources hosted by a resource server, the user must first obtain an access token from his BCA server with which he is registered. Thus, he provides his fresh biometric traits (through the sensor) together with his identity. These two bits of information are then refined by the sensor and sent to the BCA server for attestation. The BCA server queries the Database for the stored template associated with the user ID and compares it with the received one. If the templates are close enough the user will have a higher confidence level, otherwise, he will have a lower confidence level. If the obtained confidence level is lower than a predefined threshold, the user is rejected, otherwise, the BCA server generates an access token with the obtained confidence level of the user.

After receiving the user access token, the resource server decrypts it by using the blockchain and check the confidence level of this user. If the confidence level is higher enough, based on its local security policy, the resource server provides the requested resource to the user, otherwise, the user request is rejected.

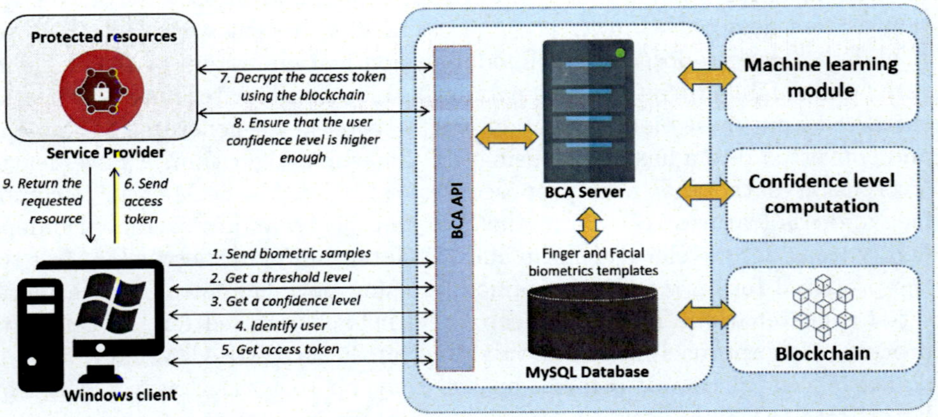

Fig. 2. High level architecture of the proposed system.

3.1 Biometrics Acquisition

The proposed multimodal biometric system integrates Tow main biometrics: fingerprint, face, and two other features: age, and gender. The fingerprint is the most successful and popular pattern that has been used for individuals identification and verification [11]. Fingerprints are unique and do not change in time. Their uniqueness is identified by the ridge's structures on the inner surface of a finger or a thumb. The ridges have unique local patterns, called minutiae, which have been widely used by forensic experts to match two fingerprints [2,21]. Ridge ending and ridge bifurcation are the most used patterns by automatic fingerprint recognition systems. A ridge ending refers to the point where a ridge ends [3,11], while a ridge bifurcation is a point where the ridge diverges into branch ridges [11]. Grayscale image, phase image, skeleton image, and minutiae are the commonly used fingerprint representation schemes in most fingerprint recognition systems [2]. In our system, The overall process of capturing the finger sample from the user takes four scans of the finger, then, the minutiae data is extracted from the Fingerprint Image Data (FID) into a template called Fingerprint Minutiae Data (FMD). The FMD is used for comparison within the system. The main reason for choosing FMD is that the original FID cannot be retrieved from the FMD as it is a one-way process.

The Facial Biometric is also known as the most distinctive key attributes for biometric authentication due to their uniqueness and robustness [13]. This technology is usually based on measurement of the facial features like mouth, eyes, nose, lips, and the face structure [13]. In this context, several techniques can be used to extract relevant features from the face image like colour analysis and neural network. In this paper, we have used the robust and fast technique Luxand FaceSDK[1] to handle the facial biometrics extraction. Luxand FaceSDK is cross-platform face detection and recognition library that provides the coordinates of over 70 facial feature points including eyes, mouth, eyebrows, nose and face contours [5]. During the enrolment phase, an image of the user is taken, and the minutiae data is extracted using Luxand SDK into a template that will be saved along with the user finger template. The facial template cannot be reversed and can only be used for comparison.

As additional features, age and gender are extracted from the submitted facial image and analysed by using the Digital Persona SDK, which returns a confidence level. For instance, age result would be 'Male: 96.9999% and Female: 3.0001%'. Then, the age result is compared against the user's gender from the database and normalised in order to be consumed by the machine learning algorithm. The finger and facial templates generated in the enrolment phase are saved in a MySQL database, while the age and gender data do not need to be stored as they can be extracted on the fly. Along with those features, some other information related to the user is also saved like his identifier, name and privileges. All communications to the MySQL database are done through an ASP Web 2.0 API that was developed throughout this work.

3.2 Biometrics Verification and Normalisation

During the verification phase, the extracted facial and finger samples of each user are used for matching with those stored in MySQL database during the enrolment phase. The obtained similarity results are tested against a set of predefined thresholds. If the similarity values are greater than the predefined thresholds, then the comparison process returns the Boolean value "true", otherwise, it returns "false". In this step, the results from the verification and matching processes are different, some provide Boolean outputs depending on thresholds like the finger and facial biometrics matching, while other results are provided as percentages like the age and gender identification. Therefore, the obtained results should be normalised before they can be used by the machine learning module (see Fig. 3). For that, the collected results for age and gender features are also examined against thresholds, if their values are greater than the thresholds then, the result is "true", otherwise, it is "false".

3.3 Machine Learning and Confidence Level

The use of multimodal biometrics needs that the outcomes from multiple sources are combined to produce one result. Then the obtained result is used to figure out whether the acquired biometrics data represent a legitimate user or not. A

[1] https://www.luxand.com/facesdk/.

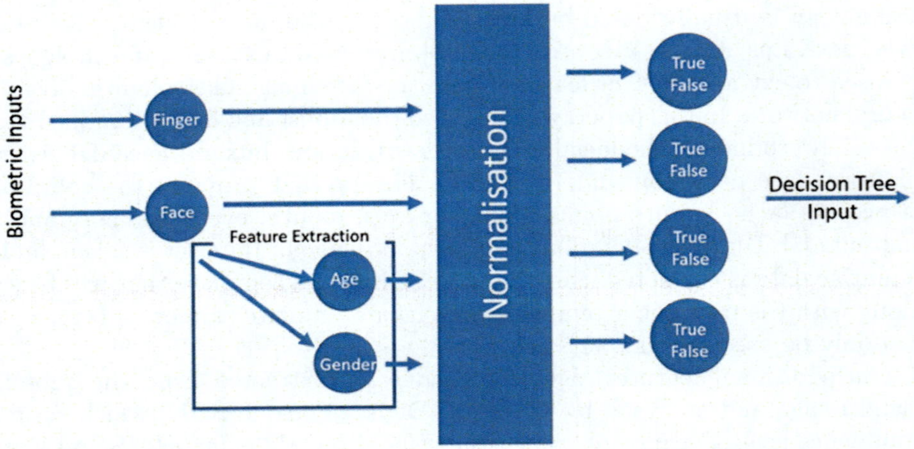

Fig. 3. Normalisation diagram.

variety of methods are available for the integration, however, in the proposed system, we will use a decision- level fusion method by using the supervised learning algorithms Decision Tree (DT) to integrate the normalised results from the four modalities and drive the confidence level related to the user. DT is a powerful and attractive approach for classification and prediction. Unlike other supervised learning algorithms, the DT has the ability to understand the given inputs and return a valuable result within a short space of time. In addition, it does not need extensive learning period compared to other methods like Neural Networks (NNs). The structure of a decision tree begins with a root node which branches out to children nodes or decision nodes, each node represents an input for the decision tree. Each children node has leaf nodes (or terminal nodes) that are the values for each of those inputs. DT predicts the value of a target variable by learning simple decision rules inferred from the data features.

The DT decision process is separated into two main steps. The first step is the training phase, where the DT is constructed and learned its training data to understand how to interpret the inputs. In the second step, the normalised inputs will be fed into the decision tree to drive a decision or a confidence percentage. This value represents the confidence level associated with the user, which will be used to update the user's confidence by adjusting the obtained confidence value directly via a connection to MySQL database. The value of this attribute is then used to produce the confidence level of the target user and decide whether to give him access to the protected resources or not. If the user does not obtain the required confidence level, he cannot access the protected resources hosted by the webserver. The required confidence level is defined by the resource server based on its local security policy. For the decision process, the biometrics were weighted as follow, finger and facial samples are weighted at 40% and age and gender are weighted at 10% (Fig. 4).

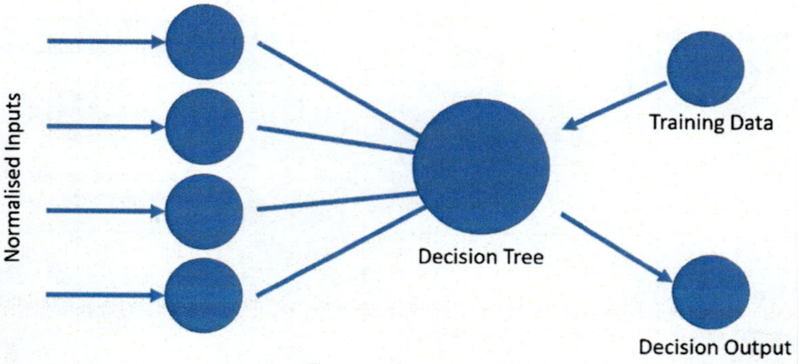

Fig. 4. Decision tree architecture.

3.4 Blockchain

The emergence of Bitcoin has highlighted the benefits of applying blockchain to the areas of identity management due to its decentralised, fault-tolerant and transparent structure that can ensure trust among different parties without relying on specific trusted, central authorities [8]. The blockchain is an encrypted ledger that is distributed and replicated among the nodes of a peer-to-peer network. It contains a linear sequence of chained blocks that can generate trust without external trusted authority. This makes it difficult to compromise the integrity of their records without being identified by the entire network, and render massive data breaches very difficult, if not theoretically impossible [4,8]. All these characteristics were contributed to the rise of many promising and innovative blockchain-based identity management solutions [14].

In this work, the blockchain consists of a number of participating resource servers and BCA servers and it is used as a public shared ledger to store user details and key data in the form of transactions. The key data stored in transactions is used by the BCA servers and untrusted resource servers to encrypt/decrypt the users 'access tokens and prove their authenticity. Each transaction contains the user's ID, timestamp, Key, start date, end date, and previous block hash. A new transaction is created and added to the chain when a new user is enrolled in the system. The BCA server does the mining for the block on creation to avoid clients having to do intensive processing to preserve the user experience. The resource server can use the blockchain to decrypt the access token sent by the user and check if his confidence level is higher than the predefined threshold.

4 Experimental Analysis

In this section, we present the experiments carried out over the proposed identity management system in order to demonstrate its effectiveness and reliability.

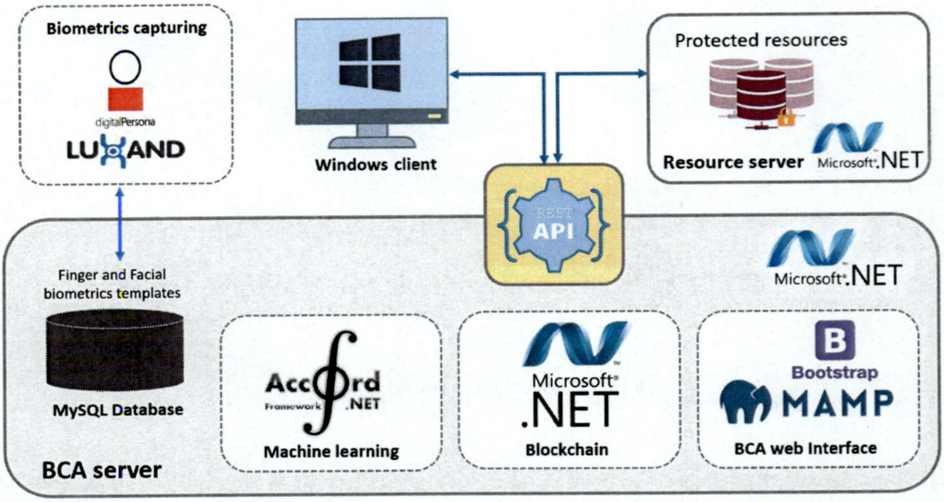

Fig. 5. Implemented technologies for the simulation experiments.

4.1 Experiment Setup

As shown in Fig. 5, the simulation experiments were performed on a client/server environment based on the Microsoft. NET technology, where each entity is deployed in a separate VM. The overall process of capturing the finger samples from the users is performed using the Fingerprint Reader Software "DigitalPersona 4500", while the facial samples have been captured using "Luxand SDK" library. Then, the templates generated from the enrolment phase are stored in MySQL database, where a BLOB field is created for each type of templates. All communications to the database were done through an ASP Web 2.0 API that was developed in this work. The machine learning component was implemented using the Accord Framework for .NET . This framework allows a smooth implementation of the DT learning algorithm on the BCA Server compared to PyTorch. Unlike other learning algorithms, DT does not require intensive training and make quick decisions, which is very important for the performance of the authentication system. The blockchain has been implemented using Microsoft. NET framework.

In this work, a user-friendly GUI has been added to provide a dashboard that can be used by administrators for controlling users, viewing analytical data, and managing predefined thresholds on the BCA system (see Fig. 6). The GUI has been implemented using the Bootstrap framework[2] which provides a quick and customised design of more professional web interface with HTML5.

[2] https://getbootstrap.com/.

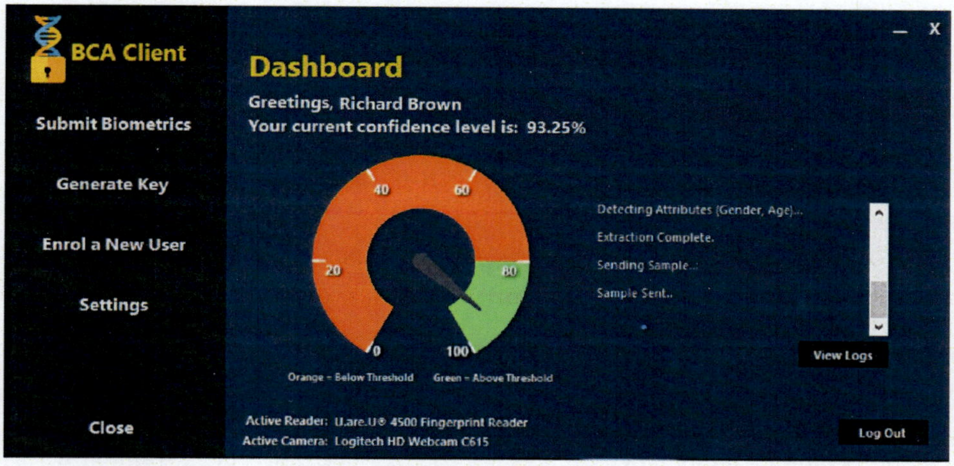

Fig. 6. BCA server dashboard.

4.2 Discussion and Results

Thresholds Values and Accuracy. For testing the efficiency and performance of the proposed authentication system, several experiments have been carried out to identify the suitable thresholds values that can offer a balance between the system performance and accuracy. Therefore, the thresholds values for the fingerprint and facial biometrics are set using the recommended SDK threshold values, which provides numerical values of the threshold for setting the False Positive Identification Rates (FPIRs) and False Acceptance Rate (FAR). FAR (also referred to as "Type II error") is the rate of frauds that were incorrectly identified over the total examined samples [25]. FAC is a critical biometric security error as it provides illegal users access to the system. Thus, it must be decreased to the minimum possible.

FPIR (also referred to as "Type I error") is the ratio of the test cases that are classified above a threshold "T" (True) over the total tested samples. The threshold "T" is used to classify samples to be either a correct (true or positive) or false (negative). If the sample is below a threshold "T" then it is classified false (negative), otherwise, it is classified true (positive). Table 1 shows the thresholds for the fingerprint biometric and their relationship with the FPIRs.

In our system, we select the threshold value 21474, which gives FPIR of .001% along with an expected number of FP of 1 in 100,000 identifications. The rate of FPI can be reduced more, however, this can increase the authentication problems for legitimate users. For example, if their fingers are sweaty, greasy, or slightly damaged, this will prevent them from accessing the system due to the authentication failures.

For the facial thresholds, the False Acceptance Rate (FAR), when the system incorrectly identifying an unauthorized person, depends on the threshold value and the total memory limit set on the capture. The higher the memory limit,

Table 1. Finger Thresholds and their Relation to FP Identification Rates

Thresholds "T"	Corresponding FPIRs	Expected number of FP identifications	Numeric value of the threshold "T"
.001*maxint	.1%	1 in 1,000	2147483
.0001*maxint	.01%	1 in 10,000	214748
.00001*maxint	.001%	1 in 100,000	21474
1.0e-6*maxint	.0001%	1 in 1,000.000	2147

the higher the false acceptance rate. FAR is also considered the most serious of biometric security errors as it may give impostors access to the system. Table 2 shows the relationship between the thresholds, the memory limit, and the FAR. In our system, the memory limit is set to 1024 MO for a facial template with a threshold of 0.992 and a FAR around 0.0002%, which is considered acceptable because it is used in combination with other biometrics. With the proposed thresholds the authentication system achieved high accuracy values ranging from 0.99% to 100%, with FAR of 0.0002% for facial biometric and FPIR of 0.001% for fingerprint.

Table 2. Facial Thresholds and their relationship with FARs and Memory Limits

Thresholds	Memory limits (MO)					
	350	700	1750	3500	5250	7500
0.992,000	0.000,041	0.000,114	0.000,703	0.001,287	0.001,938	0.002,475
0.993,141	0.000,035	0.000,104	0.000,519	0.001,099	0.001,744	0.002,010
0.994,283	0.000,031	0.000,095	0.000,462	0.000,882	0.001,377	0.001,574
0.995,424	0.000,013	0.000,054	0.000,304	0.000,646	0.000,953	0.001,212
0.996,566	0.000,013	0.000,045	0.000,211	0.000,458	0.000,700	0.000,812
0.997,707	0.000,009	0.000,038	0.000,146	0.000,314	0.000,435	0.000,566
0.988,849	0.000,006	0.000,006	0.000,066	0.000,161	0.000,276	0.000,327
0.999,990	0.000,000	0.000,003	0.000,000	0.000,003	0.000,013	0.000,022

Confidence Level. Several experiments were performed on the proposed authentication system by considering a real-world case in which the right permissions and identity of six users have been checked. The biometrics matching results are processed by the trained learning algorithm in order to provide the overall confidence level of the user at each authentication transaction. The overall threshold for the confidence level is set to 80%. This value means that at one of the less weighted features (Ager or gender) is not true. Table 3 presents the data used for training the DT learning algorithm.

Table 3. Training data for the DT learning algorithm

Finger	Face	Gender	Age	Confidence Level
True	True	True	True	100.00%
True	True	True	False	80.00%
True	True	False	True	80.00%
True	True	False	False	70.00%
True	False	True	True	60.00%
True	False	True	False	50.00%
True	False	False	True	50.00%
True	False	False	False	40.00%
False	True	True	True	60.00%
False	True	True	False	50.00%
False	True	False	True	50.00%
False	False	True	True	20.00%
False	False	False	True	10.00%
False	False	False	False	0.00%

The graph in Fig. 7 shows the evolution of confidence level values over time for one user. During this period of time, the user sent its biometric samples to the authentication system in more than 100 transactions, with different biometric

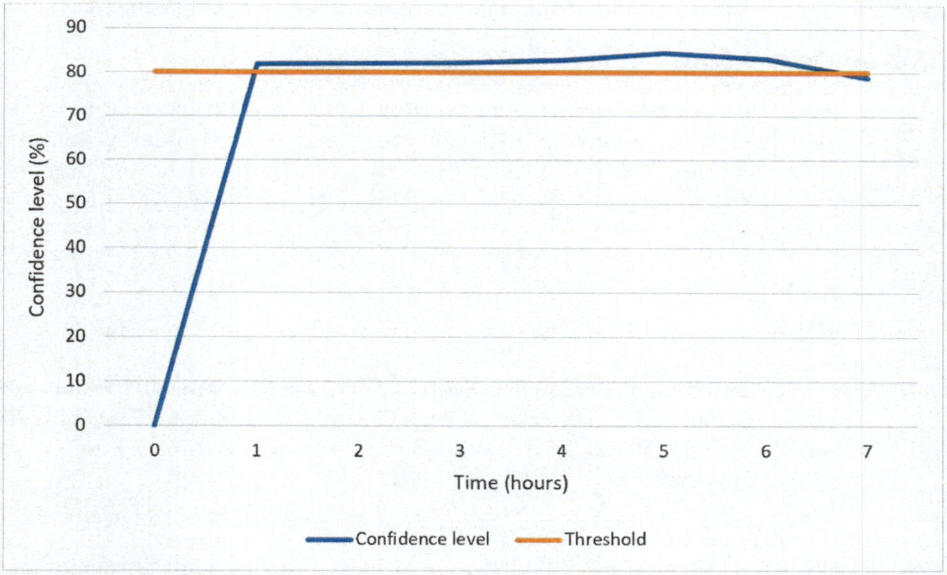

Fig. 7. Confidence level values over time for one user.

samples of this user. From the obtained results, it is noticed that the confidence level values of the user are changing as expected, where the confidence level of this user stay over the threshold (from 82% to 86%) for all good samples and has been dropped below the threshold (78.6%) with bad biometrics samples. The same observations have been achieved for all users.

5 Conclusion

In this paper, we proposed a multimodal authentication system using fingerprint and face biometrics, with age and gender features. The proposed scheme employs the DT learning algorithm to compute the user's confidence levels based on the submitted biometrics. The later is then used by the proposed system to authenticate the users. It should be higher than a predefined threshold in order the user can have access to the system. The effectiveness of the proposed system has been justified using a real-world case in which the right permissions and identity of six users have been checked, with a set of more than 100 biometric samples for each user. The samples are classified from bad to good samples.

The experiments results showed the system behaved as expected, where the good samples obtained higher confidence values and the bad samples obtained lower confidence levels. However, more experiments are needed to confirm the efficiency of the proposed approach, thus, we intend to extend this work with more experiments on large data sates from real-world as well as testing the robustness of this system against different security attacks. We also intend to further reduce the time for biometric submission by fully automating this process and minimise the user interactions, which makes the proposed system more suitable for real-time applications where computation speed is crucial.

Acknowledgement.

This project has received funding from the European Union's Horizon 2020 research and innovation programme under grant agreement no. 786698. This work reflects authors' view and Agency is not responsible for any use that may be made of the information it contains.

References

1. Azzini, A., Marrara, S., Sassi, R., Scotti, F.: A fuzzy approach to multimodal biometric continuous authentication. Fuzzy Optim. Decis. Making **7**(3), 243 (2008)
2. Bansal, R., Sehgal, P., Bedi, P.: Minutiae extraction from fingerprint images-a review. arXiv preprint arXiv:1201.1422 (2011)
3. Clarke, N., Furnell, S.: A composite user authentication architecture for mobile devices. J. Inf. Warfare **5**(2), 11–29 (2006)
4. Dunphy, P., Petitcolas, F.A.: A first look at identity management schemes on the blockchain. IEEE Secur. Priv. **16**(4), 20–29 (2018)
5. FaceSDK, L.: Luxand facesdk – overview. shorturl.at/tQZ23 Accessed 21 Jul 2020

6. Furnell, S.M.: From passwords to biometrics: In pursuit of a panacea. In: Camp, O., Weippl, E., Bidan, C., Aïmeur, E. (eds.) Information Systems Security and Privacy International Conference on Information Systems Security and Privacy, pp. 3–15. Springer, Cham (2015)
7. Hammad, M., Liu, Y., Wang, K.: Multimodal biometric authentication systems using convolution neural network based on different level fusion of ecg and fingerprint. IEEE Access **7**, 26527–26542 (2018)
8. Jacobovitz, O.: Blockchain for identity management. The Lynne and William Frankel Center for Computer Science Department of Computer Science. Ben-Gurion University, Beer Sheva (2016)
9. Jagadiswary, D., Saraswady, D.: Biometric authentication using fused multimodal biometric. Procedia Comput. Sci. **85**, 109–116 (2016)
10. Joseph, T., Kalaiselvan, S., Aswathy, S., Radhakrishnan, R., Shamna, A.: A multimodal biometric authentication scheme based on feature fusion for improving security in cloud environment. J. Ambient Intell. Human Comput. 1–9 (2020)
11. Kataria, A.N., Adhyaru, D.M., Sharma, A.K., Zaveri, T.H.: A survey of automated biometric authentication techniques. In: 2013 Nirma University International Conference on Engineering (NUiCONE), pp. 1–6. IEEE (2013)
12. Kaur, H., Khanna, P.: Remote multimodal biometric authentication using visual cryptography. In: Chaudhuri, B., Nakagawa, M., Khanna, P., Kumar, S. (eds.) Proceedings of 3rd International Conference on Computer Vision and Image Processing, pp. 13–25. Springer, Singapore (2020)
13. Liu, X., Zhu, S., Wang, W., Liu, J.: Alde: privacy risk analysis of analytics libraries in the android ecosystem. In: Deng, R., Weng, J., Ren, K., Yegneswaran, V. (eds.) International Conference on Security and Privacy in Communication Systems, pp. 655–672. Springer, Cham (2016)
14. Liu, Y., Zhao, Z., Guo, G., Wang, X., Tan, Z., Wang, S.: An identity management system based on blockchain. In: 2017 15th Annual Conference on Privacy, Security and Trust (PST), pp. 44–4409. IEEE (2017)
15. Lumini, A., Nanni, L.: Overview of the combination of biometric matchers. Inf. Fusion **33**, 71–85 (2017)
16. Nadheen, M.F., Poornima, S.: Feature level fusion in multimodal biometric authentication system. Int. J. Comput. Appl. **69**(18), 36–40 (2013)
17. Oloyede, M.O., Hancke, G.P.: Unimodal and multimodal biometric sensing systems: a review. IEEE Access **4**, 7532–7555 (2016)
18. Peng, J., Abd El-Latif, A.A., Li, Q., Niu, X.: Multimodal biometric authentication based on score level fusion of finger biometrics. Optik **125**(23), 6891–6897 (2014)
19. Saevanee, H., Clarke, N., Furnell, S., Biscione, V.: Continuous user authentication using multi-modal biometrics. Comput. Secur. **53**, 234–246 (2015)
20. Sarier, N.D.: Multimodal biometric identity based encryption. Future Gener. Comput. Syst. **80**, 112–125 (2018)
21. Shen, W., Eshera, M.: Feature extraction in fingerprint images. In: Ratha, N., Bolle, R. (eds.) Automatic Fingerprint Recognition Systems, pp. 145–181. Springer, New York (2004)
22. Tharwat, A., Ibrahim, A.F., Ali, H.A.: Multimodal biometric authentication algorithm using ear and finger knuckle images. In: 2012 Seventh International Conference on Computer Engineering & Systems (ICCES), pp. 176–179. IEEE (2012)

23. Toli, C.A., Preneel, B.: A survey on multimodal biometrics and the protection of their templates. In: Camenisch, J., Fischer-Hübner, S., Hansen, M. (eds.) IFIP International Summer School on Privacy and Identity Management, pp. 169–184. Springer, Cham (2014)
24. Weforum: World economic forum annual meeting. https://www.weforum.org/events/world-economic-forum-annual-meeting-2020. Accessed 17 Jul 2020
25. Weik, M.H.: Identification and verification. Comput. Sci. Commun. Dict 745–745 (2000)

User Attribution Through Keystroke Dynamics-Based Author Age Estimation

Ioannis Tsimperidis[1](✉), Shahin Rostami[2], Kevin Wilson[2], and Vasilios Katos[2]

[1] Democritus University of Thrace, Komotini, Greece
itsimper@ee.duth.gr
[2] Bournemouth University, Poole, UK

Abstract. Keystroke dynamics analysis has often been used in user authentication. In this work, it is used to classify users according to their age. The authors have extended their previous research in which they managed to identify the age group that a user belongs to with an accuracy of 66.1%. The main changes made were the use of a larger dataset, which resulted from a new volunteer recording phase, the exploitation of more keystroke dynamics features, and the use of a procedure for selecting those features that can best distinguish users according to their age. Five machine learning models were used for the classification, and their performance in relation to the number of features involved was tested. As a result of these changes in the research method, an improvement in the performance of the proposed system has been achieved. The accuracy of the improved system is 89.7%.

Keywords: Keystroke dynamics dataset · User age classification · Feature selection · Information gain · RBFN · AUC · Digital evidence

1 Introduction

In the original study [1] the authors proposed a system to collect information about an attacker who, by stealing the identity of a legitimate user, had managed to enter a computer system illegally. This information included inherent characteristics, such as gender and age, and also acquired characteristics, such as educational level and computer experience. This information was then used in a forensic investigation to find the guilty party. The research was focused on the task of trying to classify unknown users into age groups, by exploiting data that came from the way a user uses the keyboard. Specifically, by using 120 digram latencies, one of the most widely used keystroke dynamics features, a success rate of 66.1% was achieved in predicting which age group the user belonged to. It is noted that users were divided into four age groups, giving a random prediction rate of 25%.

It is clear that any such system, that can detect the age or other characteristics of a totally unknown user, can be used as an investigation tool in any computer crime. For example, if someone tries to break into a computer system, tries to mislead an unsuspecting user, or carries out some cyberbullying, it is likely that the malicious user will use a

B. Ghita and S. Shiaeles (Eds.): INC 2020, LNNS 180, pp. 47–61, 2021.
https://doi.org/10.1007/978-3-030-64758-2_4

keyboard - in which case, the system can create a profile giving valuable information to digital forensics agents. Indeed, the need for such a system, or for something similar, is becoming increasingly urgent due to the amount of cybercrime in the world today [2].

In the previous research, some goals were set for further work, including expanding the available dataset, by recording more users during the daily use of their computers, and using multiple classifiers in parallel, the results from which would be summarised with the use of Dempster-Shafer theory [3]. To achieve these goals, a three step plan was developed. First, a new user recording phase took place; second, more features of the keystroke dynamics were utilized; and third, additional classifiers were used.

The objective of the project has remained the same, which is to recognize certain user traits in the most universal and most economical way possible, whilst ensuring the privacy of the users and minimizing their harassment.

The rest of the paper is organized as follows. In the next section, the relevant literature in this and similar fields is summarised. The following section describes the experimental methods used: namely the data acquisition, the selection of suitable features, and the evaluation procedure. Then the results from the use of five well-known machine learning models, namely support vector machine (SVM), simple logistic (SL), Bayes classifier (NB), Bayesian network classifier (BNC), and radial basis function network (RBFN), are presented. Finally, the paper concludes by listing the conclusions and plans for further research.

2 Background

Classifying computer users according to their characteristics could be useful in a variety of applications. For example, in automated translation, due to the fact that many languages have some system of grammatical gender, a more successful translation might result when the gender of the user who is typing is known. Another example is targeted advertising, where characteristics such as gender, age and educational level are important parameters for determining a user's interests. Yet another example is in strengthening user authentication, where the more user characteristics that can be used for comparison, the greater the possibility that the user can be correctly identified. Human-computer interaction applications can also benefit from the ability to recognise user characteristics, as it enables them to modify their interface, and display user specific messages. Because of these and other applications, the classification of computer users has captured the attention of many researchers.

For example, Zhang et al. [4] exploited text features typed by users under different conditions, such as when writing a new message or responding to someone else's message, and tried to classify them according to their gender and age, dividing them into two age groups. With the help of an LSTM network, they managed to achieve a successful prediction rate of 90.8% in gender classification, and 82.3% in age classification. In another paper, Culotta et al. [5] collected data from Twitter user profiles and enriched them with demographic data from an audience measurement company, thus creating their dataset. Using features derived from the association of users with other users, such as who the user follows, and text features from the tweets that they wrote, and employing a distantly-supervised regression model, they were able to determine the user gender with an F1 value of 0.87. They were also able to determine the nationality, out of four different choices, with an F1 value of 0.81, and also the political orientation, out of two political parties, with an F1 value of 0.74.

One approach to classifying users is to use features extracted from facial pictures, as in the work of Zhang et al. [6], who extracted features from five points of the face, and then cropped the image in different ways. For each resulting image, they used a convolutional neural network to perform gender classification, where they achieved a success rate of about 91.7%, and smile classification, where they achieved a maximum success rate of about 89.3%. Another attempt is that of Chikkala et al. [7], who used data from four facial picture databases and divided users into six age groups. Their method was based on the third order four pixel pattern and achieved a 96% success rate in each of the databases they used, surpassing performance over all other active methods.

Most user classification methods, such as those mentioned above, are based on features that were extracted from users' photos, videos they appear in, text that they typed, websites that they visited, or a profile that they maintain in a social network. Of course, each of these methods serves the purpose for which it was proposed. However, with regard to the search for forensic evidence in cases where a cybercrime has been committed, most, if not all, of these methods are considered inadequate.

There are several reasons for this. One is that a malicious user will usually try to conceal or misrepresent their identity, and so will not perform any attacks through their genuine social network account, Internet service, or computing system, and so there will be no available picture of the attacker. Consequently, methods that perform user classification through facial images, or the examination of website traffic, or through user profile data, cannot be used for this purpose. Another reason is that methods that examine text written by the user, are based on the extraction of the required features from words and other parts of speech, such as digrams and trigrams, from a particular language. This means that these methods have serious limitations when they are used to examine text from a different language.

An alternative approach, which overcomes the aforementioned problems, is to use keystroke dynamics, a field of computer science that studies how users type. The advantage of keystroke dynamics information is that it uses features from the simplest and most common form of communication between Internet users, which is text [8]. Users write emails, send instant messages, make searches in search engines, upload posts, and communicate much more frequently with text than any other method of communication, such as videoconferencing. Another advantage of using keystroke dynamics information is that no special equipment is required for their operation, except for the common QWERTY keyboard. The advantages include the non-disturbance of users as data can be collected during their daily use of the computer without requiring additional actions on their part, and independence from the typed language since the features are not related to words in a particular language.

In the past, keystroke dynamics information has been used primarily to authenticate users in order to replace or enhance user authentication by passwords.

Salem and Obaidat [9] proposed an authentication system for Android mobile devices using temporal features such as digram latencies, and non-temporal features such as on-screen pressure, finger positioning, etc. They created an application for recording volunteer's actions and as well as that data, they also used and existing dataset for their experiments. Various classifiers were tested, and MLP, with an EER of 0.9%, proved to be the one with the best performance. In another paper, Saini et al. [10] attempted to authenticate users of portable devices, regardless of their body posture when they use them. They recorded data from mobile phones with users sitting, walking, or relaxing. In the processing, the random forest and kNN classifiers were used, achieving an optimal EER of 4.3%.

As has already been mentioned, most of the research on keystroke dynamics has focused on authenticating users, with only a small percentage being oriented to other applications, such as Kolakowska's work [11], which attempts to recognize the emotional state of a keyboard user. A small amount of the research is concerned with classifying users according to some of their characteristics, such as the research of Tsimperidis et al. [12], in which the authors collected 242 logfiles from volunteers during daily use of their computers, used a combination of an RBFN classifier and a boosting algorithm, and managed to predict correctly the educational level of a user, among five options, with accuracy of 86.8%. In another paper, Brizan et al. [13] used data from 350 volunteers who typed a short piece of text, extracted textual and keystroke dynamics features, and achieved recognition of a user's mother tongue, gender, and handedness, at rates higher than those of random selection.

Regarding the classification of users according to their age with the help of keystroke dynamics, which is the subject of this research, there are some interesting studies. Buriro et al. [14] tried to investigate the possibility of estimating, among other things, the age of a user who types a PIN/password between 4 and 16 digits in length, on a smart mobile device. They collected their data from 150 volunteers on a specific device and defined 3 age groups. They extracted temporal keystroke dynamics features and used Naïve Bayes, SVM, Random Forest, MLP, and Deep Neural Network for classification. The best results came from Random Forest (RF), which had an accuracy of 87.9%. Random Forest was also the most successful classifier amongst 7 others, in the work of Roy et al. [15]. They conducted their study to protect young people from unknown threats coming from the Internet and therefore divided users into two classes, children and adults. They used three fixed text datasets from 11 to 14 keystrokes and exploited keystroke durations and digram latencies. Finally, using an Ant Colony Optimization (ACO) technique they achieved an accuracy of 92.2%. Pentel [16] divided the users into 6 groups, gathered data from more than 7,000 users, each of which was recorded for about 320 keystrokes, extracted 134 keystroke dynamics features in total, and reached an accuracy of 61.6% using Random Forest.

It is understood that a number of researches have aimed at classifying users based on one or some of their characteristics taking advantage of various types of data, such as face images and posts on social networks. However, only a very small portion of them use data derived from keystroke dynamics. This research is one of the few in user classification through keystroke dynamics and as far as we know the only one that focuses on the exploitation of results in digital forensics.

3 Method

The methodology consists of three consecutive phases. In the first phase, free-text data was collected from volunteers who agreed to participate in the experiment. In the second phase, a feature selection algorithm was used to sort the features according to the information that they contain. In the third phase, an attempt was made to determine the previously unknown age of a user by training and hyperparameter-tuning five well-known machine learning algorithms, namely SVM, Simple Logistic, Naïve Bayes, Bayesian Network, and RBFN.

3.1 Keystroke Dynamics Dataset

It was stated that one of the ways of improving the results of the previous research was to extend the available dataset. It was decided that the acquisition of data should be done in a way that interferes as little as possible with the daily use of the computer by the users. For this reason, the keylogger was designed to record actions on the keyboard, in any application, without causing any harassment to the user.

Although the research did not intend to capture the text written by a user, it was technically possible to reconstruct it from the data that was recorded. For this reason, guarantees were given to the volunteers who participated, in order to safeguard their sensitive or personal data, such as passwords, credit card numbers, or messages to third parties. Each volunteer was given a signed consent form by the researchers, stating that the recorded data would be encrypted, remain exclusively in their possession, and would not be shared with others in any way. It also stated that only the keystroke dynamics features would be studied, from which it would not be possible to reconstruct the original text. In addition, volunteers were not only made aware of the potential dangers, they were also given the ability to run the keylogger only when they wanted to, so that they could choose which data was recorded. Finally, they were allowed to oversee the data recorded, and decide at any point in the process whether or not they wished to hand it over to the researchers.

Each keyboard action made by the volunteers was recorded in the logfiles, as shown below:

78,#2017-11-14#,56861220,"dn"
78,#2017-11-14#,56861368,"up"
65,#2017-11-14#,56861502,"dn"
73,#2017-11-14#,56861728,"dn"
65,#2017-11-14#,56861742,"up"
73,#2017-11-14#,56861883,"up"

Each logfile entry consists of four parts separated by commas and corresponds to a key press or release action. The first part is the virtual key code of the key used (from 1 to 255); the second part, delimited by the sharp character (#), is the date on which the action took place; the third part is the exact time at which the action took place, as an integer denoting the ms that have passed since the beginning of the day (12:00 am); and finally the fourth part shows the type of action, with "dn" representing a key press, and "up" representing a key release.

With these additional measures, and using the software developed for this purpose in the previous study, the researchers conducted a second phase of data collection from volunteers who did not participate in the first phase. The second recording phase lasted 8.5 months, from 24/10/2017 to 09/07/2018, and 43 volunteers were selected, thus increasing the number of participants to 118, so that the demographics of the created dataset reflected those of the world population, such as ensuring that the number of males is approximately equal to the number of females, and that the number of right-handed users is about 90% of the sample [17], and most important for the present study, to have satisfactory representation of all age groups.

Table 1 shows the comparison between the initial dataset (first phase of volunteer recording) and the extended dataset (first and second phases of volunteer recording).

Table 1. Comparison between initial and extended dataset.

	Initial dataset		Extended dataset	
Age group	Number of files	Percentage	Number of files	Percentage
18–25	32	13.4%	96	24.8%
26–35	102	42.7%	129	33.3%
36–45	90	37.6%	117	30.3%
46+	15	6.3%	45	11.6%
Total	239		387	

As can be seen from Table 1, the expansion of the dataset led to a more even distribution across the age groups, as the logfiles from "18–25" and "46+" age groups, which were the least common in the initial dataset, increased in number so that their share of the overall dataset almost doubled in percentage terms.

Each of 387 logfiles is between 170 KB and 271 KB in size and contains data relating to between 2,800 and 4,500 keyboard actions. This variation in the size of the logfiles is due to two things: the fact that the keylogger was designed to record data of a certain size in bytes, and therefore, depending on the time of the day the volunteer was recorded, and depending on the keys used, the number of recorded keys could have a difference of ±5%. The other fact is that, as stated in the consent form, no volunteer was obliged to complete the recording process, which sometimes created files of smaller than normal size. Eventually, it was decided that only files exceeding a certain size threshold size would be accepted.

3.2 Feature Extraction and Feature Selection

Keystroke dynamics encompass a large number of features, which can be divided into two categories: temporal and non-temporal. The temporal features are the most widely used, and they include keystroke durations and digram latencies. Other features in the same category are the trigram, tetragram, and general n-gram latencies; the duration of pauses during typing; and the typing rate (words per unit of time). The non-temporal features include the percentage use of duplicate keys ("Shift", "Ctrl", digits, etc.); the way in which the typing errors are corrected ("Delete", "Backspace"); and the time of the day the user is typing.

Much of the research involving keystroke dynamics only makes use of a small number of the available features, with most researchers only using some of the keystroke durations and one or more of the digram latencies (down-down, down-up, up-down, and up-up). In the first phase of this research, the authors made use of 120 down-down digram latencies, which were selected according to their incidence.

In this extension to that phase of the research, the intention is to use more keystroke dynamics features and to evaluate them according to the amount of information that they provide for classifying users according to their age. However, the features examined will include those found in most researches, namely the keystroke durations and down-down digram latencies. There are a large number of these, since $n^2 + n$ features can be extracted from a keyboard with n keys. Most companies use the PC keyboard with 104 keys as a de facto standard and therefore the number of extracted features can be as large as 10,920.

This large number of features can lead to systems with high time complexity and therefore a procedure must be followed to reduce their number. This process, which is called feature selection, must identify those features which are most capable of distinguishing users according to their age. One way of doing this is to calculate the information gain (IG) of each feature f, which is the measure that illustrates the ability of that feature to reduce the entropy of a system x. It is expressed as:

$$IG(x, f) = H(x) - H(x|f) \tag{1}$$

The entropy $H(x)$ of the system x is given by:

$$H(x) = -\sum_{i=1}^{m} P(x_i) \cdot \ln P(x_i) \tag{2}$$

In Eq. (2), m is the length of vector x, which in the classification problem is the number of classes, and $P(x_i)$ is the probability of class x_i. In this study there are 4 classes and therefore the entropy of the system is 1.312. The term $H(x|f)$ is calculated by dividing the dataset into groups according to the value of the particular feature f. Then, the entropy of each group is calculated and $H(x|f)$ is given by:

$$H(x|f) = \frac{1}{N} \sum_{j=1}^{k} n_j \cdot H(x_j) \tag{3}$$

where N is the number of instances of the initial dataset, k is the number of groups that the initial dataset was divided into, n_j is the number of instances of the j-th group, and $H(x_j)$ is the entropy of the j-th group, which can be calculated from Eq. (2).

This procedure is also described in the work of Osanaiye et al. [18] and, if applied to every extracted feature in the age classification problem, it will produce a list with the amount of information that every feature carries. A list of 15 features with the highest

IG is shown in Table 2, where the keystroke durations are represented with one number (such as "69", the first in the list) and digram latencies are represented with two numbers, separated by a dash (such as "65–32", the second in the list).

Table 2. Keystroke dynamics features with the highest IG in age classification.

#	Feat.	IG	#	Feat.	IG	#	Feat.	IG
1	69	0.1457	6	32	0.0781	11	86	0.0659
2	65–32	0.1377	7	39	0.0746	12	84–79	0.0637
3	79	0.1006	8	87	0.0741	13	87–32	0.0620
4	65	0.0802	9	83	0.0721	14	70	0.0618
5	68	0.0791	10	89	0.0689	15	88	0.0592

As can be seen in Table 2, keystroke durations appear to play a more important role than digram latencies in user classification based on their age.

3.3 Experimental Procedure and Validation of Models

The feature selection procedure used showed that more than 90% of the features extracted contain zero *IG*, so they may be excluded from user classification, resulting in a huge reduction in time complexity with a minimal or no reduction in accuracy.

The other features, those with non-zero *IG*, were all used to predict the unknown age of a user. Various classifiers were tested for this purpose, several of which showed very low success rates, such as Random Forest, C4.5, k-Nearest Neighbors, Random Tree, and OneR, while others had a prohibitively long training time, such as the MLP, which was the classifier used in the previous research of the authors. The five models that presented high accuracy and low time complexity were SVM, SL, NB, BNC, and RBFN, and therefore the experimental process continued with them.

The model validation stage is to ensure that the implementations of the models are correct and work as they should. There are many techniques that can be utilized to verify a model and several of them were adopted to validate the five models.

First, to assess the performance of the models fairly, we use the 10-folds cross-validation method. This divides the data into 10 disjoint parts, uses 9 of them for training and the remaining one for testing, in a round-robin fashion. In this study where the dataset consists of 387 log files, each fold will consist of 38 or 39 files.

Second, in order to evaluate the effectiveness of the feature selection procedure, the F-score was also used as a combined measurement of precision and recall, because accuracy alone cannot give the full picture of the overall performance of a model when classes are imbalanced, and also because the F-score is a measurement of how balanced the prediction is between classes.

Finally, in order to assess the ranking ability of the classifiers, use is made of the receiver operating characteristic (ROC) curve, which shows recall as a function of the probability of a false negative, which is equivalent to 1 – precision. The area under the

ROC curve (AUC) or ROC index [19] was used. The ROC curve is limited to the interval [0, 1] in both dimensions, thus the AUC varies between 0 and 1.

4 Experiments and Results

For each of the five models (SVM, SL, NB, BNC, and RBFN) several experiments were conducted to find the classifier parameters that implement the system with the optimal performance for different sets of features. The first criterion was the performance with the highest accuracy (Acc.), with the second being the one with the lowest time complexity (TBM - Time to Build Model), followed by the highest Area Under the ROC Curve (AUC) and the highest F-score (F1).

Experiments were done with various sets of features in order to evaluate the performance of these models. These involved using different numbers of keystroke dynamics features, starting with the first 100 features according to their IG value and finishing with 700 features, in steps of 100.

The best performance of SVM for different number of features, along with the optimal C value, is shown in Table 3.

Table 3. The performance of SVM over different number of features

# of feats.	Statistical values				Classifier parameters	
	Acc.	TBM	AUC	F1	C	Kernel
100	65.9%	0.06	0.796	0.642	0.8	Polykernel
200	71.1%	0.12	0.826	0.706	2.0	Polykernel
300	72.9%	0.11	0.845	0.722	1.0	Polykernel
400	73.6%	0.17	0.853	0.732	2.0	Polykernel
500	74.4%	0.25	0.861	0.741	4.0	Polykernel
600	75.5%	0.27	0.864	0.752	4.0	Polykernel
700	74.2%	0.19	0.851	0.732	0.5	Polykernel

Several conclusions can be drawn from Table 3. First, the accuracy and the F-score, in each different set of features, exceeds the corresponding measures of the same classifier in the previous study, which was 56.5% and 0.545 respectively. Second, as expected, time complexity is too low, even when several features are involved. Third, the polynomial kernel works better than the other kernel types.

Similarly, Table 4 shows the performance of SL and the corresponding optimal values for the last iteration of LogitBoost, over the seven different feature sets, if no new error minimum has been reached.

From Table 4 it follows that the Simple Logistic model shows better accuracy and F-score than the corresponding classifier in the prior study, which were 55.7% and 0.552, respectively.

Table 4. The performance of SL over different number of features

# of Feats.	Statistical values				Classifier parameters	
	Acc.	TBM	AUC	F1	Last iteration	Weight trimming
100	63.1%	0.56	0.826	0.625	50	95%
200	67.4%	1.86	0.859	0.672	50	100%
300	71.1%	1.58	0.874	0.706	60	85%
400	72.9%	5.11	0.879	0.726	150	95%
500	73.6%	7.59	0.879	0.734	200	95%
600	73.1%	10.61	0.879	0.727	80	100%
700	71.6%	9.86	0.877	0.712	50	100%

The results for the NB classifier are in Table 5.

Table 5. The performance of NB over different number of features

# of Feats.	Acc.	TBM	AUC	F1
100	62.3%	0.03	0.829	0.620
200	67.2%	0.06	0.835	0.664
300	67.7%	0.03	0.837	0.670
400	68.2%	0.06	0.839	0.673
500	66.9%	0.06	0.835	0.660
600	66.9%	0.02	0.831	0.660
700	66.9%	0.02	0.830	0.660

Two findings from Table 5 are that the Naïve Bayes model shows improved accuracy and F-score in each set of features, compared to the previous research, which produced the values 50.2% and 0.488 respectively, and that, as expected, the time complexity of the model is very low.

The best results for BNC are shown in Table 6, which also presents the corresponding optimal initial count on each feature set for estimating the probability tables and the optimal maximum number of parents of each node in Bayes network.

Table 6 reveals the seemingly contradictory result that the BNC presents higher time complexity when the number of features involved is smaller, which is due to the different settings of the classifier that led to its best performance in each case. The BNC model was not examined in the previous work and no direct comparison can be made.

Table 6. The performance of BNC over different number of features

# of feats.	Statistical values				Classifier parameters	
	Acc.	TBM	AUC	F1	Initial count	Max number of parents
100	66.2%	0.38	0.836	0.660	0.10	5
200	67.4%	0.62	0.858	0.674	0.10	3
300	67.7%	1.36	0.865	0.676	0.20	3
400	68.7%	0.05	0.875	0.685	0.01	1
500	69.0%	0.05	0.878	0.689	0.02	1
600	70.0%	0.06	0.886	0.699	0.01	1
700	69.8%	0.08	0.886	0.697	0.01	1

Finally, the results from the optimal configuration of RBFN in terms of the number of clusters for K-Means and the minimum standard deviation for the clusters yielding the best performance, are presented in Table 7.

Table 7. The performance of RBFN over different number of features

# of feats.	Statistical values				Classifier parameters	
	Acc.	TBM	AUC	F1	# of clusters	Min std dev
100	82.7%	1.30	0.917	0.827	130	1.1
200	86.6%	2.31	0.942	0.866	110	1.1
300	88.9%	2.70	0.950	0.889	110	1.4
400	89.7%	3.55	0.960	0.897	120	1.2
500	89.2%	4.22	0.949	0.891	110	1.4
600	89.2%	5.03	0.953	0.892	110	1.2
700	89.2%	5.67	0.954	0.892	110	1.2

As can be seen from Table 7, the RBFN presents the best performance for each set of features at similar values of the classifier's parameters, namely a value between 110 and 130 for the number of clusters for K-Means, and a value between 1.1 and 1.4 for the minimum standard deviation for the clusters. The RBFN model was also not considered in the previous study.

4.1 Evaluation and Comparison of Results

The best performance of each of the examined models is illustrated in Fig. 1.

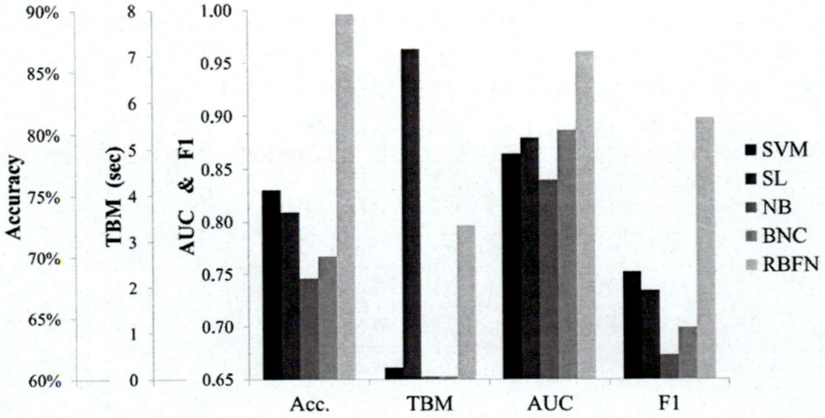

Fig. 1. Comparison of the best performances of the five models

As can be seen, the RBFN model outperforms all other models in terms of accuracy, AUC, and F-score. SVM follows as second in accuracy and F-score, but also lags behind SL and BNC in AUC. NB ranks last out of the five models in performance, but is the fastest of all, along with the BNC.

The accuracy of the five models, against the number of keystroke dynamics features used, is shown in Fig. 2.

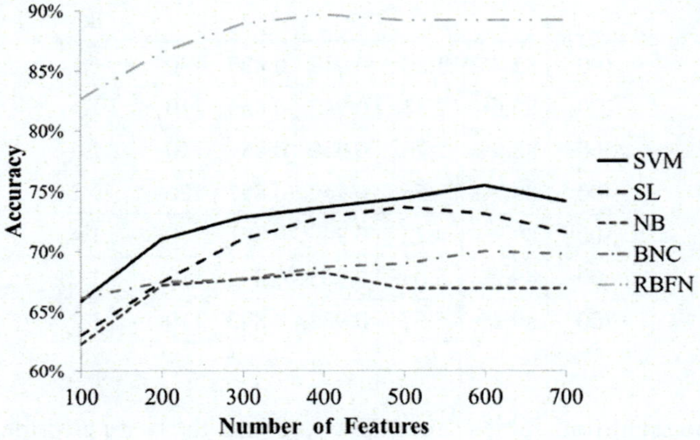

Fig. 2. Accuracy of five models on various feature sets

It can be seen from Fig. 2 that RBFN presents the highest accuracy against all other models, regardless of the feature set, whereas NB presents the lowest. Also, the performance of BNC has the least dependence on the number of features used, as its accuracy is 68% ± 2% regardless of size of the feature set. One last point is that each of the models achieves the highest precision using less than 700 features. This is a strong indication that a large number of features is not necessarily required to achieve high accuracy, and perhaps of more importance is the use of appropriate features.

5 Future Research Directions

Research into the creation of systems that can recognize some characteristics of an unknown user by the way the keyboard is used, has already shown significant results. However, it can be extended in various directions in order to improve the reliability of those systems.

One such direction would be to carry out a new phase of volunteer recording, which will enlarge the existing keystroke dynamics dataset. One of the objectives of this new phase would be to record users of as many different ethnicities as possible, so that the dataset is enriched with logfiles of many different mother tongues, each of which will be adequately represented. The purpose of the study will be to search for indications as to whether the keystroke dynamics features are independent of the language being typed and, if they are not, the nature of the dependency, and with how much accuracy an unknown user's mother tongue can be predicted. Alternatively, existing datasets such as the one created by Clarkson University [20] can be used.

Apart from the age and the mother tongue of an individual, there are some other characteristics, either acquired or inherent, that can be used to form the user's profile. Among them is gender, for which there are already several studies with very good results. Also, the user's handedness, and their educational level. Creating an integrated system that can extract from an unknown user an accurate set of characteristics, with a very high percentage of success, is the ultimate goal of this research. Indeed, if this is achieved with as little keystroke dynamics data as possible, e.g. a few tens or hundreds of keystrokes, then it could be used as a valuable forensic tool when a digital crime has been committed.

Another interesting direction is to investigate how much a user can modify the way that they type, so as to mislead a recognition system into producing false results. It leads on to the interesting question of whether a user's typing pattern is something superficial, which can be easily disguised, or whether it is more deep rooted, implying that it will inevitably be detected by a suitable system, thus enabling the user profile to be extracted correctly after a certain number of keystrokes.

Finally, another possible avenue of further research is the use of alternative features of keystroke dynamics. As has already been mentioned, in addition to keystroke durations and digram latencies which are widely used, there are many other keystroke dynamics features that could provide much more information to aid in identifying user characteristics. For example, quantitative and qualitative differentiation in pauses during typing may contain some information to enable characteristics classification. Moreover, the preference shown by some users for using a specific key when there are two similar options, such as the "Shift" key, may also contain extra information. The study of the possibility of classifying users with some hitherto underused features, and their combination with those that are widely used, is another possible extension of this work.

6 Conclusion

The evolution of technology and the transfer of a large part of human activity to the Internet, including communication, work, entertainment, and education, has also resulted in increased crime in this area. A crime that has very different quality characteristics to that of the physical world, as everyone can hide behind a device and a public network.

This research attempts to identify the age of an unknown user as a part of his/her profile, amongst other characteristics such as gender, handedness, and educational level, according to the way he/she is typing, in order to facilitate a forensic investigation. For this reason, keystroke dynamics features, namely keystroke durations and digram latencies, were extracted from a dataset of 387 logfiles created for this purpose. The process followed was the initial extraction of features, then the selection of some of them according to the amount of information they include which would aid age classification, and finally the use of some well-known machine learning models for the classification of a user into one of four age groups.

The results showed that the age group a user belongs can be predicted with an accuracy of almost 90%, far exceeding the random prediction accuracy which is 25% (one of four possible choices) and the accuracy level reported in the authors' previous research, which was 66.1%. Therefore, it seems possible to identify some characteristics of an unknown user with high accuracy rates and thereby implement systems that can create profiles of malicious users.

The ability to identify gender, age, and other characteristics of a user may have useful applications beyond forensic investigation. For example, in the field of targeted advertising, where filters are used to determine whether or not a user is exposed to advertising material, there will be much more data at their disposal, resulting in more efficient and less disturbing advertising. Some other applications relate to user convenience, such as automatically filling in certain fields of a form when creating an account, or the suggestions given to a user for things like visiting websites, or joining groups, depending on his/her interests. Finally, an equally important application is that of alerting unsuspecting users to the possibility of becoming a victim of an online fraud. For example, it could be used to warn a young person, who thought that they were communicating with someone of a similar age to themselves, that in fact the person that they were messaging belonged to the age group "46+ ", with a very high probability.

Acknowledgements. This work has been partially supported by IDEAL-CITIES; a European Union's Horizon 2020 research and innovation staff exchange programme (RISE) under the Marie Skłodowska-Curie grant agreement No 778229.

References

1. Tsimperidis, I., Rostami, S., Katos, V.: Age detection through keystroke dynamics from user authentication failures. Int. J. Digit. Crime Forensics **9**(1), 1–16 (2017)
2. Mendoza, D.K.O.: The vulnerability of cyberspace - the cyber crime. J. Forensic Sci. Crim. Invest. **2**(1), 1–8 (2017)
3. Jirousek, R., Shenoy, P.P.: A new definition of entropy of belief functions in the Dempster-Shafer theory. Int. J. Approximate Reasoning **92**(1), 49–65 (2018)

4. Zhang, D., Li, S., Wang, H., Zhou G.: User classification with multiple textual perspectives. In: Proceedings of 26th International Conference on Computational Linguistics, pp. 2112–2121, The COLING 2016 Organizing Committee, Osaka, Japan (2016)
5. Culotta, A., Ravi, N.K., Cutler, J.: Predicting Twitter user demographics using distant supervision from website traffic data. J. Artif. Intell. Res. **55**, 389–408 (2016)
6. Zhang, K., Tan, L., Li, Z., Qiao, Y.: Gender and smile classification using deep convolutional neural networks. In: Proceedings of 2016 IEEE Conference on Computer Vision and Pattern Recognition Workshops, pp. 34–38. IEEE, Las Vegas, NV, USA (2016)
7. Chikkala, R., Edara, S., Bhima, P.: Human facial image age group classification based on third order four pixel pattern (TOFP) of wavelet image. Int. Arab J. Inf. Technol. **16**(1), 30–40 (2019)
8. Walther, J.B., Van Der Heide, B., Ramirez, A.J., Burgoon, J., Pena, J.: Interpersonal and hyperpersonal dimensions of computer-mediated communication. In: Sundar, S.S. (ed.) The Handbook of Psychology and Communication Technology, pp. 3–22. John Wiley & Sons, Inc. (2015)
9. Salem, A., Obaidat, M.S.: A novel security scheme for behavioral authentication systems based on keystroke dynamics. Secur. Priv. **2**(2), 1–11 (2019)
10. Saini, B.S., Kaur, N., Bhatia, K.S.: Position independent mobile user authentication using keystroke dynamics. In: Pandey, B., Khamparia, A. (eds.) Hidden Link Prediction in Stochastic Social Networks, pp. 64–78. IGI Global (2019)
11. Kolakowska, A.: Recognizing emotions on the basis of keystroke dynamics. In: Proceedings of 8th International Conference on Human System Interaction, pp. 75–80. IEEE, Warsaw, Poland (2015)
12. Tsimperidis, I., Yoo, P.D., Taha, K., Mylonas, A., Katos, V.: R^2BN: An adaptive model for keystroke-dynamics-based educational level classification. IEEE Trans. Cybern. **50**(2), 525–535 (2020)
13. Brizan, D.G., Goodkind, A., Koch, P., Balagani, K., Phoha, V.V., Rosenberg, A.: Utilizing linguistically enhanced keystroke dynamics to predict typist cognition and demographics. Int. J. Hum Comput Stud. **82**, 57–68 (2015)
14. Buriro, A., Akhtar, Z., Crispo, B., Del Frari, F.: Age, gender and operating-hand estimation on smart mobile devices. In: Proceedings of 2016 International Conference of the Biometrics Special Interest Group, pp. 273–280. IEEE, Darmstadt, Germany (2016)
15. Roy, S., Roy, R., Sinha, D.D.: ACO-random forest approach to protect the kids from internet threats through keystroke. Int. J. Eng. Technol. **9**(3S), 279–285 (2017)
16. Pentel, A.: Predicting user age by keystroke dynamics. In: Silhavy, R. (ed.) CSOC2018 2018. AISC, vol. 764, pp. 336–343. Springer, Cham (2019). https://doi.org/10.1007/978-3-319-91189-2_33
17. Guadalupe, T., Mathias, S.R., vanErp, T.G.M., et al.: Human subcortical brain asymmetries in 15,847 people worldwide reveal effects of age and sex. Brain Imaging and Behav. **11**(5), 1497–1514 (2017)
18. Osanaiye, O., Cai, H., Choo, K.-K., Dehghantanha, A., Xu, Z., Dlodlo, M.: Ensemble-based multi-filter feature selection method for DDoS detection in cloud computing. EURASIP J. Wireless Commun. Networking **2016**(1), 1–10 (2016). https://doi.org/10.1186/s13638-016-0623-3
19. Hu, N.: Using receiver operating characteristic (ROC) analysis to evaluate information-based decision-making. In: Khosrow-Pour, M. (ed.) Advanced Methodologies and Technologies in Business Operations and Management, pp. 764–776. IGI Global (2019)
20. Clarkson University Keystroke Dataset. https://citer.clarkson.edu/research-resources/biometric-dataset-collections-2/clarkson-university-keystroke-dataset/. Accessed 30 Aug 2020

propagation condition into account, the most commonly expected behavior of a device is to connect to the 2.4 GHz band of an active legitimate AP if it gets disconnected from its AP.

This is the biggest disadvantage of this type of attack since the positioning of the attacker must be such, to provide signal power advantage against the original AP. In most cases this is not feasible, especially for indoor environments. Of course, in specific geometrical setups, the attacker may gain advantage using highly directional antennas. Furthermore, in some devices the internal algorithm may force the device to connect to an AP that has better capabilities than 2.4 GHz, so the device may for example return to an 802.11ac AP operating at the 5 GHz band, even if the rogue AP provides a stronger signal at 2.4 GHz.

2.3 Disassociation Attack

A disassociation attack is similar to a deauthentication attack. The purpose of a disassociation packet is to inform the AP that the station is leaving the current cell to roam to a different one within the same BSS. A station could be authenticated but not associated. According to [14], a disassociation attack takes more time to be executed since upon receiving a disassociation frame, the STAs will first issue a deauthentication frame to the AP, and then go through a complete authentication and association/re-association cycle.

2.4 Jamming Attack

According to a classification presented in [15], there are five types of jammers. The constant jammer, the intermittent jammer, the reactive jammer, the adaptive jammer and the intelligent jammer. The first four types exploit the shared nature of the wireless medium and can be regarded as wireless physical-layer jammer attacks. The intelligent jammer exploits the vulnerabilities in upper layer protocols in 802.11 medium. A novel approach in reactive jammers [16] has been carried out by Mathy Vanhoef and Frank Piessens, in which they used commodity hardware to create a reactive jammer. They demonstrated that their method could lead to a high percentage of success when small distances are involved.

The disadvantage of jamming attacks is that the efficiency is inversely proportional to the relevant distances between the devices participating in the link, and if not reactive jamming is employed, then the attack could also disrupt the communication of other users leading to a not stealthy approach.

3 Testbed and Results

A channel switch announcement MiTM attack should be more effective compared to the attacks referred in the previous sections, since it is less offensive, leading to a stealthier attack. The attack can be performed through two processes. The first process creates a fake beacon indicating that the AP is moving to a different channel due to non-Wi-Fi transmissions. Using this method all clients which are associated with the corresponding BSSID will disconnect and will send a probe request to the relevant channel communicated by the beacon. Table 1 shows the relevant channel switch announcement information.

Table 1. Beacon Frame in IEEE, "Std 802.11h" [10]

Order	Information	Notes
11	Country	The Country information element shall be present when dot11MultiDomainCapabilityEnabled is true or dot11SpectrumManagementRequired is true
14	Power constraint	Power Constraint element shall be present if dot11SpectrumManagementRequired is true
15	Channel switch announcement	Channel Switch Announcement element may be present if dot11SpectrumManagementRequired is true
16	Quiet	Quiet element may be present if dot11SpectrumManagementRequired is true
17	IBSS DFS	IBSS DFS element shall be present if dot11SpectrumManagementRequired is true in an IBSS
18	TPC Report	TPC Report element shall be present if dot11SpectrumManagementRequired is true

If an attacker aims to perform a MiTM attack to a specific device, and not to all the devices associated to the BSS, then the channel switch announcement action frame should be used. In our setup we performed the attack by using the action frame to silently manipulate the target to connect to the rogue AP.

3.1 Scenario Description

To confirm that an Evil Twin attack is possible even when the rogue AP signal power is lower than the legitimate one, two different scenarios have been tested.

In the first scenario, a dual band AP was deployed as the legitimate one and another rogue AP was deployed at the 5GHz band. The positioning of the target was such that the received signal strength from the rogue AP was higher than the signal power from the legitimate AP. The configuration of the APs and the target are shown in the table below. In the second scenario the target was receiving a higher signal strength from the legitimate AP when compared to the rogue AP. Both scenarios have been executed using the process: the legitimate and rogue APs were deployed in a building floor in different rooms, and subsequently the target, a smart device, roamed the building floor in such a way in order to receive different signal strength levels from the two APs. A non-symmetrical radio path loss was introduced between the target device and the APs through the in-between walls and different geometrical distances involved (Table 2).

Table 2. APs characteristics

Role	5 GHz band	2.4 GHz band
Legitimate AP	802.11ac channel 36	802.11n channel 1
Rogue AP	802.11ac channel 44	n/a
Target	802.11a/ac supported	802.11b/g/n supported

3.2 Rogue AP's Received Signal Strength Higher Than Legitimate AP

Initially the target is connected to the legitimate AP. In order to deploy the attack, the rogue AP is deployed. After that, specially crafted packets are injected, sending the rogue Channel Switch Announcement (CSA). The transmitted CSA appears coming from the legitimate AP informing the target that is going to operate on a different channel. The packet structure consists of five fields, as shown in Fig. 1.

The first field is the ElemendID which was set to 37, indicating a channel switch announcement. The second field denotes the length of the following fields. The channel switch mode indicates any transmission restrictions until a channel switch. This field could be either 1 or 0. A channel switch mode set to 1 means that the STA in a BSS to which the frame containing the element is addressed shall transmit no further frames within the BSS until the scheduled channel switch. If the switch mode is set to 0, the STA should transmit frames before the scheduled channel switch.

The new channel number field is set to the number of the channel to which the STA is moving. In this case the crafted packet channel number should have the operating channel number of the rogue AP.

The channel switch count field is set to the number of target beacon transmission times (TBTTs), until the STA sending the Channel Switch Announcement element switches to the new channel, or is set to 0. A value of 1 indicates that the switch will occur immediately before the next TBTT. A value of 0 indicates that the switch will occur at any time after the frame containing the element is transmitted. During this experiment, a value of 0 was used, since such value informs the device that must leave the BSS immediately. After receiving the packets, the target should roam to the rogue AP (Fig. 3).

After deploying the first scenario, the smart device of Table 3 was used as a target. Device was placed at a position around the floor, receiving the signal strengths from the two APs as indicated in Table 4.

As expected, the received signal strength from the 5 GHz band of the legitimate AP was lower than the 2.4 GHz signals from the same AP. During this experiment, the target

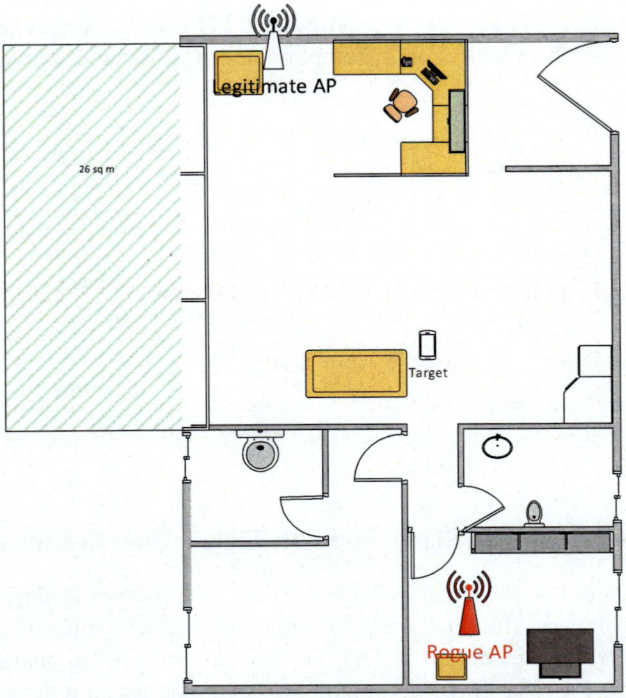

Fig. 3. Scenario where rogue AP received signal strength is higher

Table 3. Table with devices characteristics

Vendor	Model	Wi-Fi Chipset
LG	Nexus 5X	QualComm QCA6174A

Table 4. RSSI values acquired using Wi-Fi Analyzer android app at target phone.

Device	From legitimate AP@2.4 GHz (dBm)	From legitimate AP@5 GHz (dBm)	From rogue AP@5 GHz (dBm)
Nexus 5x	−63	−71	−46

device moved closer to the Rogue AP transmitting at the 5 GHz band, leading to stronger received signals from the rogue AP. At this stage, the action frame was issued, and the target device moved to the rogue AP confirming the validity of the attack.

3.3 Legitimate AP's Received Signal Strength Higher Than Rogue AP

In this scenario the position of the target device was changed to vary again the received signal strengths. As mentioned in the previous section, the target device was positioned in such a way that the received signal strength of the legitimate AP is higher than the rogue AP.

After altering the setup, the MiTM attack is deployed identically to the first setup to verify that the manipulation is not affected by the received signal strength.

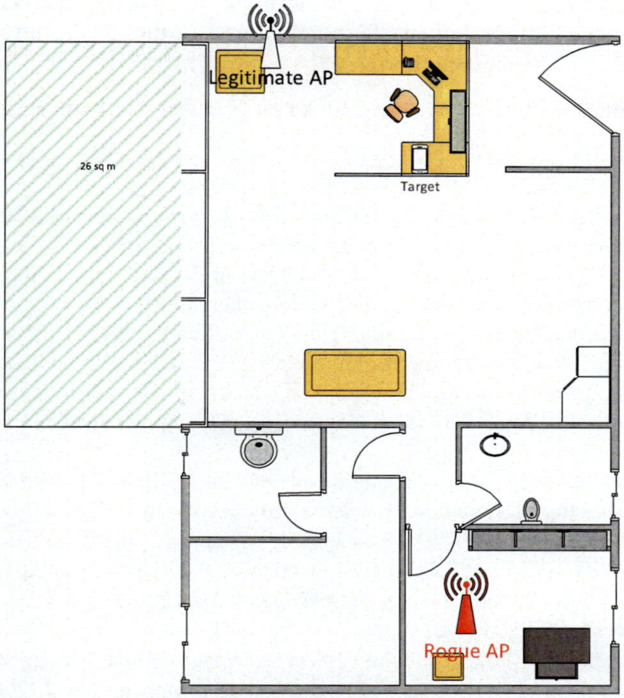

Fig. 4. Scenario where legitimate AP received signal strength is higher

Table 5. RSSI values acquired using Wi-Fi Analyzer android app at target phone

Device	From legitimate AP@2.4 GHz (dBm)	From legitimate AP@5 GHz (dBm)	From rogue AP@5 GHz (dBm)
Nexus 5x	−45	−52	−67

Similarly, to the previous scenario, the received signal strength from the 5 GHz band of the legitimate AP was lower than the 2.4 GHz signals received from the same AP. During this experiment, the target device moved closer to the legitimate AP, as also denoted in Fig. 4, leading to stronger received signals from the legitimate AP. At this stage, the action frame was again issued, and the target device connected to the rogue AP that was providing a weaker signal, as denoted by Table 5, confirming the validity of the attack.

4 Conclusions

Results obtained from the measured scenarios suggest that it is feasible to perform a successful MiTM attack even if the rogue AP provides a weaker signal to the target, compared to the legitimate AP. This means that the rogue AP can easily compete with the legitimate AP in a MiTM attack. Furthermore, the usage of directional antennas at the rogue AP means that relatively far away targets or targets withing buildings can be susceptible to a MiTM attack with the CSA approach.

References

1. LWN Page. https://lwn.net/2001/1011/a/wlan-security.php3. Released 2001
2. Fluhrer, S., Mantin, I., Shamir, A.: Weaknesses in the key scheduling algorithm of RC4. In: Vaudenay, S., Youssef, A.M. (eds.) Selected Areas in Cryptography. SAC 2001. Lecture Notes in Computer Science, vol 2259. Springer, Heidelberg (2001)
3. darkAudax aircrack-ng page. Tutorial: How to Crack WPA/WPA2 2010/03/07
4. Hashcat Page. https://hashcat.net/forum/thread-7717.html Accessed 08 Apr 2018
5. Vanhoef, M., Piessens, F.: Key reinstallation attacks: forcing nonce reuse in WPA2. In: Proceedings of the 2017 ACM SIGSAC Conference on Computer and Communications Security (2017)
6. Stefan Viehböck blog: https://sviehb.files.wordpress.com/2011/12/viehboeck_wps.pdf. Wi-Fi Protected Setup PIN brute force vulnerability. Accessed 27 Dec 2011
7. Hostapd Page. https://w1.fi/hostapd/. hostapd: IEEE 802.11 AP, IEEE 802.1X/WPA/WPA2/EAP/RADIUS Authenticator. Accessed 12 Jan 2013
8. Esser420 Github Page, Evil Twin Framework. https://github.com/Esser420/EvilTwinFramework. Accessed 07 Nov 2017
9. John, B., Stefan, S.: 802.11 Denial-of-Service Attacks: Real Vulnerabilities and Practical Solutions. In: Proceedings of 12 USENIX Security Symposium, p. 2 (2003)
10. Könings, B., Schaub, F., Kargl, F., Dietzel, S.: Channel switch and quiet attack: new DoS attacks exploiting the 802.11 standard. In: 2009 IEEE 34th Conference on Local Computer Networks, Zurich (2009)
11. IEEE: Std 802.11h - Part 11: Wireless LAN MAC and PHY Layer specifications - Amendment 5: Spectrum and Transmit Power Management Extensions in the 5 GHz band in Europe. IEEE (2003)
12. Roney, P.: Securing Wireless Networks from ARP Cache Poisoning (2003). https://digilander.libero.it/SNHYPER/files/arppoison.pdf
13. Stavrou, S., Saunders, S.R.: Review of constitutive parameters of building materials. In: Twelfth International Conference on Antennas and Propagation (ICAP 2003). (Conf. Publ. No. 491), Exeter, UK, vol. 1, pp. 211–215 (2003). https://doi.org/10.1049/cp:20030052
14. Kolias, C., Kambourakis, G., Stavrou, A., Gritzalis, S.: Intrusion detection in 802.11 networks: empirical evaluation of threats and a public dataset. IEEE Commun. Surv. Tutor. (2015)
15. Zou, Y., Zhu, J., Wang, X., Hanzo, L.: A survey on wireless security: technical challenges, recent advances, and future trends. Proc. IEEE **104**(9), 1727–1765 (2016)
16. Vanhoef, M., Piessens, F.: Advanced Wi-Fi attacks using commodity hardware, pp. 256–265 (2014). https://doi.org/10.1145/2664243.2664260

IoT

Machine Learning Based IoT Intrusion Detection System: An MQTT Case Study (MQTT-IoT-IDS2020 Dataset)

Hanan Hindy[1(✉)], Ethan Bayne[1], Miroslav Bures[2], Robert Atkinson[3], Christos Tachtatzis[3], and Xavier Bellekens[3]

[1] Division of Cyber Security, Abertay University, Dundee, Scotland, UK
{1704847,e.bayne}@abertay.ac.uk, hananhindy@ieee.org

[2] Department of Computer Science, FEE, Czech Technical University in Prague, Prague, Czechia
miroslav.bures@fel.cvut.cz

[3] EEE Department, University of Strathclyde, Glasgow, Scotland, UK
{robert.atkinson,christos.tachtatzis,xavier.bellekens}@strath.ac.uk

Abstract. The Internet of Things (IoT) is one of the main research fields in the Cybersecurity domain. This is due to (a) the increased dependency on automated device, and (b) the inadequacy of general-purpose Intrusion Detection Systems (IDS) to be deployed for special purpose networks usage. Numerous lightweight protocols are being proposed for IoT devices communication usage. One of the distinguishable IoT machine-to-machine communication protocols is Message Queuing Telemetry Transport (MQTT) protocol. However, as per the authors best knowledge, there are no available IDS datasets that include MQTT benign or attack instances and thus, no IDS experimental results available.

In this paper, the effectiveness of six Machine Learning (ML) techniques to detect MQTT-based attacks is evaluated. Three abstraction levels of features are assessed, namely, packet-based, unidirectional flow, and bidirectional flow features. An MQTT simulated dataset is generated and used for the training and evaluation processes. The dataset is released with an open access licence to help the research community further analyse the accompanied challenges. The experimental results demonstrated the adequacy of the proposed ML models to suit MQTT-based networks IDS requirements. Moreover, the results emphasise on the importance of using flow-based features to discriminate MQTT-based attacks from benign traffic, while packet-based features are sufficient for traditional networking attacks.

Keywords: IoT · Machine Learning · MQTT · Intrusion Detection

1 Introduction

A large number of Internet of Things (IoT) devices and networks have been utilised over the past years for different usage scenarios [13]. These use-cases

B. Ghita and S. Shiaeles (Eds.): INC 2020, LNNS 180, pp. 73–84, 2021.
https://doi.org/10.1007/978-3-030-64758-2_6

include healthcare [4], smart cities [5], supply chain [1] and farming [3]. With this extended use of IoT, new protocols are being deployed [17]. One of the new prominent protocols used for machine-to-machine communication is MQTT [19].

Harsha *et al.* [8] discuss the different protocols used in various IoT networks, which include MQTT. The authors analyse the security risks associated with using MQTT. The authors results show that there are 53396 publicly available and accessible MQTT devices [8]. Their work, alongside the work by Dinculeană and Cheng [7], emphasises on the need for robust detection techniques for MQTT attacks to overcome the security vulnerabilities.

As discussed in [10], IoT Intrusion Detection Systems (IDS) have different requirements due to the uniqueness of the usage scenarios involved. IoT IDSs are required to be flexible, extendable, and built using real or simulated traffic suited for the intended usage [9]. However, publicly available IoT datasets are limited, thus limiting IoT IDS development [9].

In this manuscript, we aim at proposing and evaluating different Machine Learning (ML) based MQTT IDS. The contributions of this paper are as follows:

- Generating a novel IoT -MQTT dataset and releasing it for public consumption.
- Analysing a novel MQTT dataset which includes both benign and attack scenarios.
- Evaluating the significance of using high-level (flow-based) features to build the IDS.
- Assessing the proposed model using six different ML techniques.
- Examining the different needs of MQTT-based versus generic attacks detection, which emphasise the special setup and, thus the needs of MQTT (IoT) networks.

The remainder of this paper is organised as follows; Sect. 2 discusses the setup used for the dataset generation and provides an overview of the dataset and the extracted features. Section 3 presents the results obtained by applying different ML techniques to detect attacks. Finally, the paper is concluded in Sect. 4.

2 Dataset

This section provides a description of the dataset gathered by the MQTT sensors simulation. The dataset is published[1] in [12]. The dataset consists of five recorded scenarios; normal operation and four attack scenarios. The attacker performs four attack and each is recorded independently.

[1] https://ieee-dataport.org/open-access/mqtt-iot-ids2020-mqtt-internet-things-intrusion-detection-dataset.

The attack types are:

- Aggressive scan (Scan_A)
- User Datagram Protocol (UDP) scan (Scan_sU)
- Sparta SSH brute-force (Sparta)
- MQTT brute-force attack (MQTT_BF)

The data is acquired using tcpdump. The packets are collected by recording Ethernet traffic and then exporting to pcap files. The following tools were used as follows:

- Virtual machines are used to simulate the network devices.
- Nmap is used for the scanning attacks.
- VLC is used to simulate the camera feed stream.
- MQTT-PWN [2] is used for the MQTT brute-force attack.

Figure 1 visualises the network components. The network consists of 12 MQTT sensors, a broker, a machine to simulate camera feed, and an attacker. During normal operation, all 12 sensors send randomised messages using the "Publish" MQTT command. The length of the messages is different between sensors to simulate different usage scenarios. The messages content is randomly generated. The camera feed is simulated using VLC media player which uses UDP stream. To further simulate a realistic scenario each of the network emulators drop packets with 0.2%, 1%, and 0.13%. During the four attack scenarios recording, the background normal operation was left in action. The operating systems of the different devices are as follows; Tiny Core Linux for the sensors, Ubuntu for the camera & camera feed server, and finally, Kali Linux for the hacker.

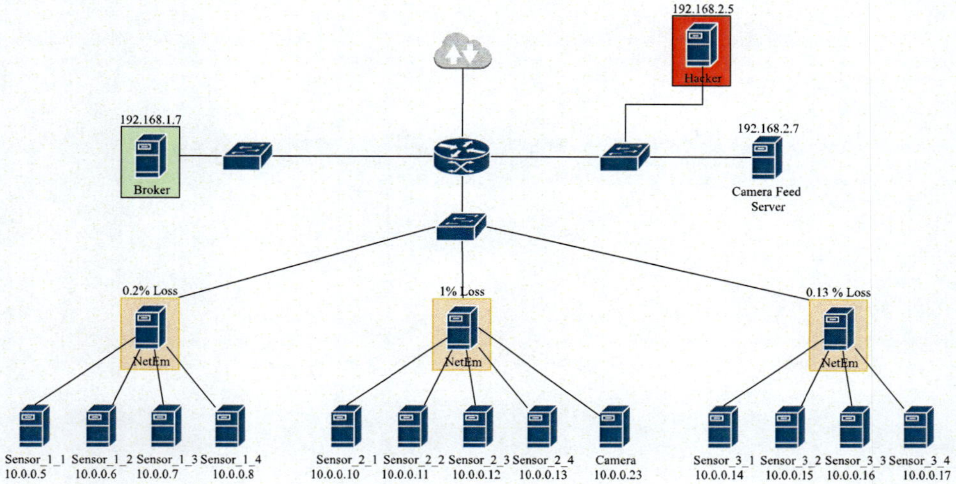

Fig. 1. MQTT network architecture [12]

The importance of this dataset is fourfold:

- The dataset simulates a realistic MQTT IoT network in a normal operation scenario.
- The dataset includes both generic networking scanning attacks, as well as, MQTT brute-force attack.
- Researchers can use this dataset to build and evaluate IoT Intrusion Detection Systems.
- The dataset is the first to include MQTT scenarios and attacks data.

The dataset is provided in its raw capture format (.pcap files), as well as processed features [12]. The features represent: (a) packet-based features, (b) Unidirectional-based features, and (c) bidirectional-based features [18]. Each feature set is used exclusively, as discussed in Sect. 3. The basic packet extracted features are listed in Table 1, fourth column. The feature list for unidirectional and bidirectional is listed in Table 1, columns five and six, respectively. It is important to note that for the bidirectional flows, some features (pointed as *) have two values—one for the forward flow and one for the backward flow. The two features are recorded and distinguished by a prefix "fwd_" for forward and "bwd_" for backward [12]. Furthermore, the distribution of instances is listed in Table 2.

In order to avoid specific features influence, the following features are dropped. These features are source and destination IP addresses, protocol, and MQTT flags. The data is split into 75% and 25% for training and testing, respectively.

Table 1. Features Description

Feature	Data type	Description	Packet	Uni-flow	Bi-flow
ip_src	Text	Source IP Address	✓	✓	✓
ip_dest	Text	Destination IP Address	✓	✓	✓
protocol	Text	Last layer protocol	✓		
ttl	Integer	Time to live	✓		
ip_len	Integer	Packet Length	✓		
ip_flag_df	Binary	Don't fragment IP flag	✓		
ip_flag_mf	Binary	More fragments IP flag	✓		
ip_flag_rb	Binary	Reserved IP flag	✓		
prt_src	Integer	Source Port	✓	✓	✓
prt_dst	Integer	Destination Port	✓	✓	✓
proto	Integer	Transport Layer protocol (TCP/UDP)		✓	✓
tcp_flag_res	Binary	Reserved TCP flag	✓		
tcp_flag_ns	Binary	Nonce sum TCP flag	✓		
tcp_flag_cwr	Binary	Congestion Window Reduced TCP flag	✓		
tcp_flag_ecn	Binary	ECN Echo TCP flag	✓		
tcp_flag_urg	Binary	Urgent TCP flag	✓		
tcp_flag_ack	Binary	Acknowledgement TCP flag	✓		
tcp_flag_push	Binary	Push TCP flag	✓		
tcp_flag_reset	Binary	Reset TCP flag	✓		
tcp_flag_syn	Binary	Synchronization TCP flag	✓		
tcp_flag_fin	Binary	Finish TCP flag	✓		
num_pkts	Integer	Number of Packets in the flow		✓	*
mean_iat	Decimal	Average inter arrival time		✓	*
std_iat	Decimal	Standard deviation of inter arrival time		✓	*
min_iat	Decimal	Minimum inter arrival time		✓	*
max_iat	Decimal	Maximum inter arrival time		✓	*
num_bytes	Integer	Number of bytes		✓	*
num_psh_flags	Integer	Number of push flag		✓	*
num_rst_flags	Integer	Number of reset flag		✓	*
num_urg_flags	Integer	Number of urgent flag		✓	*
mean_pkt_len	Decimal	Average packet length		✓	*
std_pkt_len	Decimal	Standard deviation packet length		✓	*
min_pkt_len	Decimal	Minimum packet length		✓	*
max_pkt_len	Decimal	Maximum packet length		✓	*
mqtt_messagetype	Integer	MQTT message type	✓		
mqtt_messagelength	Binary	MQTT message length	✓		
mqtt_flag_uname	Binary	User Name MQTT Flag	✓		
mqtt_flag_passwd	Binary	Password MQTT flag	✓		
mqtt_flag_retain	Binary	Will retain MQTT flag	✓		
mqtt_flag_qos	Integer	Will QoS MQTT flag	✓		
mqtt_flag_willflag	Binary	Will flag MQTT flag	✓		
mqtt_flag_clean	Binary	Clean MQTT flag	✓		
mqtt_flag_reserved	Binary	Reserved MQTT flag	✓		
is_attack	Binary	1 if the instance represents an attack, 0 otherwise	✓	✓	✓

*represented as two features in the biflow features file (forward fwd and backward bwd)

Table 2. Dataset instances distribution

File Name	pcap file size	Number of packets		Number of Uni-flow instances		Number of Uni-flow instances	
		Benign	Attack	Benign	Attack	Benign	Attack
normal	192.5 MB	1056230 (3.42%)	0	171836 (59.01%)	0	86008 (54.78%)	0
scan_A (aggressive)	16.2 MB	70768	40624 (0.13%)	11560	39797 (13.67%)	5786	19907 (12.68%)
Scan_sU (UDP)	41.3 MB	210819	22436 (0.07%)	34409	22436 (7.71%)	17230	22434 (14.29%)
sparta	3.4 GB	947177	19728943 (63.93%)	154175	28232 (9.7%)	77202	14116 (8.99%)

3 Experiments and Results

This section discusses the conducted experiments. Note that the code is available on a GitHub repository [2].

Five-fold cross validation is used to evaluate each experiment. The metrics used for evaluation are as follows [9]: (a) Overall accuracy, as defined in Eq. 1, such that True Positive (TP) represents the attack instances correctly classified, True Negative (TN) represents the benign instances correctly classified, Positive (P) represents the number of attack instances and Negative (N) represents the total number of benign instance.

$$OverallAccuracy = \frac{TP + TN}{P + N} \tag{1}$$

For each class, Precision, Recall, and F1 Score are computed as shown in Eq. 2, Eq. 3, and Eq. 4, respectively [9]. False Positive (FP) represents benign instances falsely classified as attack and False Negative (FN) represents the attack instances falsely classified as benign.

$$Precision = \frac{TP}{TP + FP} \tag{2}$$

$$Recall = \frac{TP}{TP + FN} \tag{3}$$

$$F1 = \frac{2TP}{2TP + FP + FN} \tag{4}$$

Finally, the weighted average for precision, recall, and F1 score is calculated to demonstrate the overall performance.

Six ML techniques are employed for the classification purpose. The ML techniques are: Logistic Regression (LR), Gaussian Naïve Bayes (NB) , k-Nearest Neighbours (k-NN) , Support Vector Machine (SVM) , Decision Trees (DT) and Random Forests (RF) [6,11,14–16,20,21].

[2] https://github.com/AbertayMachineLearningGroup/MQTT_ML.

Table 3 details the overall accuracy of each of the ML techniques with packet, unidirectional and bidirectional features. It can be observed the performance rise accompanying flow-based features, both unidirectional and bidirectional. This rise could further be visualised in Fig. 2.

Table 3. Overall detection accuracy

	Features		
	Packet	Unidirectional	Bidirectional
LR	78.87%	98.23%	99.44%
k-NN	69.13%	99.68%	99.9%
DT	88.55%	99.96%	99.95%
RF	65.39%	99.98%	99.97%
SVM (RBF Kernel)	77.4%	97.96%	96.61%
NB	81.15%	78%	97.55%
SVM (Linear Kernel)	66.69%	82.6%	98.5%

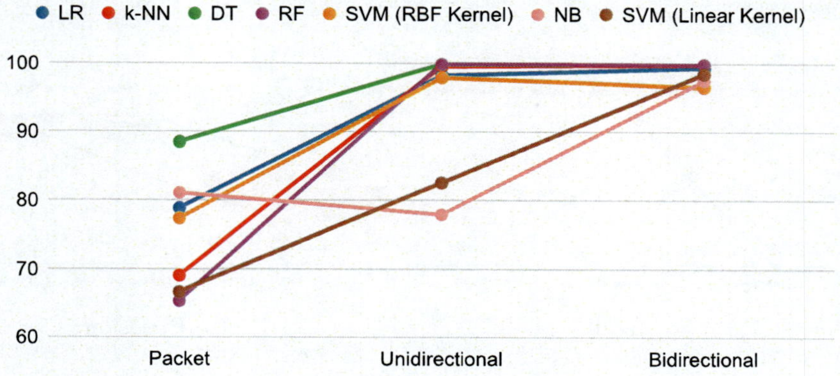

Fig. 2. Overall detection accuracy trend using different ML techniques

To further analyse the results, Table 4 and Table 5 show the detailed precision, recall, and F1-score for each of the classifiers. Each classifier is represented as a sub-table. Similar to Table 3, it is observed that the flow-based features are strongly enhance the results.

Furthermore, it is recognised that the Benign and the MQTT-BF attack are the two classes benefiting from flow features. This is reasoned by the fact that in IoT networks normal/benign operations are usually uncomplicated, due to their usage, requirements, and the nature of data of interest. Therefore, generic attacks are quite distinctive. However, MQTT-based attacks have similar characteristics to benign MQTT communication. Since MQTT-based attacks rely on

Table 4. 5-fold cross validation

	Recall			Precision			F1-score		
	Packet	Uni	Bi	Packet	Uni	Bi	Packet	Uni	Bi
LR									
Benign	0%	100%	99.02%	0%	93.33%	98.95%	0%	96.55%	98.99%
Scan_A	86.45%	70.87%	97.25%	98.39%	98.39%	97.21%	92.03%	82.39%	97.2%
Scan_sU	98.21%	98.03%	98.48%	99.34%	95.76%	100%	98.77%	96.88%	99.23%
Sparta	100%	100%	100%	98.22%	100%	100%	99.1%	100%	100%
MQTT_BF	100%	99.25%	99.58%	51.75%	99.82%	99.41%	68.2%	99.53%	99.5%
Weighted average	78.87%	98.23%	99.44%	70.4%	98.32%	99.44%	72.97%	98.14%	99.44%
k-NN									
Benign	17.43%	99.69%	99.95%	17.42%	98.85%	99.59%	17.43%	99.27%	99.77%
Scan_A	99.99%	99.97%	100%	99.99%	99.85%	99.9%	99.99%	99.91%	99.95%
Scan_sU	99.99%	99.96%	100%	99.99%	99.96%	100%	99.99%	99.96%	100%
Sparta	100%	100%	100%	100%	100%	100%	100%	100%	100%
MQTT_BF	25.84%	99.3%	99.75%	25.85%	99.82%	99.97%	25.84%	99.56%	99.86%
Weighted average	69.13%	99.68%	99.9%	69.13%	99.68%	99.9%	69.13%	99.68%	99.9%
DT									
Benign	69.29%	99.92%	99.88%	69.39%	99.92%	99.91%	69.34%	99.92%	99.9%
Scan_A	100%	100%	100%	99.98%	99.95%	99.9%	99.99%	99.97%	99.95%
Scan_sU	99.98%	99.91%	100%	100%	100%	100%	99.99%	99.96%	100%
Sparta	100%	100%	100%	100%	100%	100%	100%	100%	100%
MQTT_BF	72.56%	99.95%	99.93%	72.47%	99.95%	99.93%	72.51%	99.95%	99.93%
Weighted average	88.55%	99.96%	99.95%	88.55%	99.96%	99.95%	88.54%	99.96%	99.95%

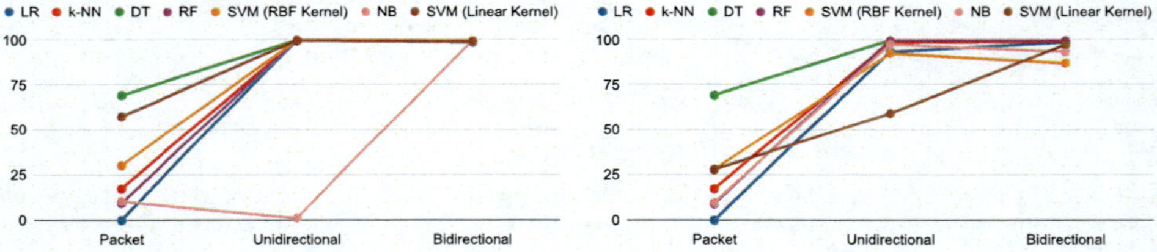

(a) Recall (b) Precision

Fig. 3. Benign class trends

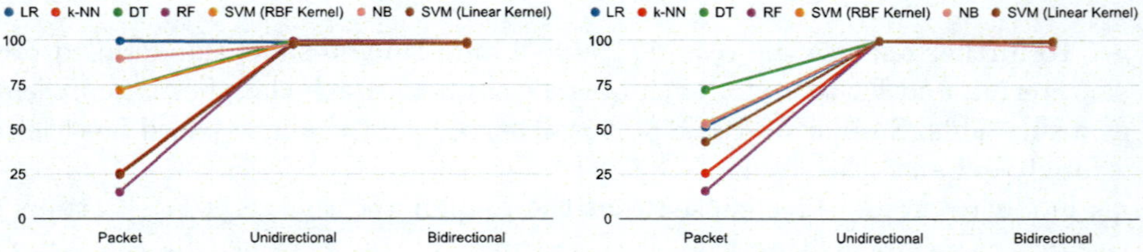

(a) Recall (b) Precision

Fig. 4. MQTT_BF class trends

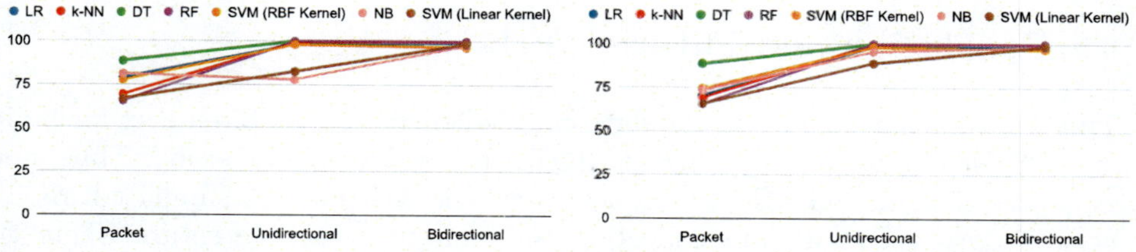

(a) Recall (b) Precision

Fig. 5. Weighted average trends

the available MQTT communication commands (i.e., publish, subscribe, etc.), it is challenging to discriminate attacks from normal operations where the same commands are used. As a result, packet-based features in all the ML techniques were not suitable for benign and MQTT-BF classification. This observation could further be observed in the trends charts for benign class, MQTT_BF class, and weighted average metrics in Fig. 3, Fig. 4 and, Fig. 5.

Table 5. 5-fold cross validation

	Recall			Precision			F1-score		
	Packet	Uni	Bi	Packet	Uni	Bi	Packet	Uni	Bi
				RF					
Benign	9.34%	99.96%	99.93%	8.99%	99.94%	99.95%	9.16%	99.95%	99.94%
Scan_A	100%	100%	100%	99.98%	99.95%	99.95%	99.99%	99.97%	99.98%
Scan_sU	99.98%	99.91%	99.96%	99.99%	100%	100%	99.99%	99.96%	99.98%
Sparta	100%	100%	100%	100%	100%	100%	100%	100%	100%
MQTT_BF	15.15%	99.96%	99.97%	15.69%	99.98%	99.96%	15.42%	99.97%	99.97%
Weighted average	65.39%	99.98%	99.97%	65.44%	99.98%	99.97%	65.41%	99.98%	99.97%
				SVM (RBF Kernel)					
Benign	30.23%	100%	100%	28.13%	92.67%	87.13%	28.8%	96.19%	93.12%
Scan_A	83.8%	70.16%	42.13%	99.99%	96.18%	99.88%	91.18%	81.13%	59.22%
Scan_sU	92.33%	99.96%	100%	99.74%	93.01%	94.34%	95.89%	96.36%	97.09%
Sparta	100%	100%	100%	91.17%	100%	100%	95.38%	100%	100%
MQTT_BF	72.42%	98.44%	98.3%	53.56%	100%	100%	59.53%	99.22%	99.14%
Weighted average	77.4%	97.96%	96.61%	74.35%	98.05%	97.02%	74.89%	97.87%	96.15%
				NB					
Benign	10.62%	1.13%	99.96%	9.9%	97.68%	93.56%	10.25%	2.24%	96.65%
Scan_A	100%	99.25%	66.41%	99.23%	18.28%	100%	99.61%	30.88%	79.81%
Scan_sU	99.52%	97.76%	100%	100%	98.79%	98.52%	99.76%	98.27%	99.25%
Sparta	99.84%	100%	100%	100%	100%	100%	99.92%	100%	100%
MQTT_BF	90.27%	97.78%	100%	53.15%	100%	97.05%	65.84%	98.88%	98.5%
Weighted average	81.15%	78%	97.55%	73.29%	95.43%	98.37%	75.99%	75.26%	97.77%
				SVM (Linear Kernel)					
Benign	57.34%	99.84%	99.26%	27.8%	58.95%	97.45%	37.38%	73.82%	98.32%
Scan_A	83.28%	68.23%	84.1%	70.42%	70.35%	93.44%	69.7%	67.5%	87.01%
Scan_sU	78.13%	60.31%	97.76%	75.8%	70.71%	93.77%	76.92%	61.91%	95.27%
Sparta	87.64%	60.37%	99.99%	97.62%	99.94%	100%	89.89%	74.61%	99.99%
MQTT_BF	24.89%	97.79%	98.71%	43.3%	99.89%	99.55%	20.84%	98.83%	99.13%
Weighted average	66.69%	82.6%	98.5%	65.42%	88.9%	98.66%	60.4%	82.42%	98.46%

4 Conclusion and Future Work

This work aims at exploring the different challenges and requirements for building IDS for IoT networks, using an MQTT network as a case study. This paper evaluates six different ML techniques as attack classifiers. A simulated MQTT network was used for data collection to simulate a real-life setup. Using the dataset raw pcap files, three features levels were extracted; packet, unidirectional, and bidirectional features. Each feature level is used independently in the experiments. The experiments highlighted that generic networking attacks are easily discriminated from normal operation due to their distinguished behaviour and patterns compared to the IoT setup. However, MQTT-based attacks are more complicated and can easily mimic benign operation.

The experimental results further demonstrated that the flow-based features are better suited to discriminate between benign and MQTT-based attacks due to their similar characteristics. The weighted average recall rose from ~75.31% for packet-based features to ~93.77% and ~98.85% for unidirectional and bidirectional flow features, respectively. While the weighted average precision rose from ~72.37% for packet-based features to ~97.19% and ~99.04% for unidirectional and bidirectional flow features. Therefore, the experiments emphasised on the special challenges faced by IoT IDS, based on their custom communication patterns. The challenges were demonstrated through the difficulties to differentiate MQTT-based attacks from normal operations.

References

1. Abdel-Basset, M., Manogaran, G., Mohamed, M.: Internet of things (IoT) and its impact on supply chain: a framework for building smart, secure and efficient systems. Future Gener. Comput. Syst. **86**, 614–628 (2018)
2. Abeles, D., Zioni, M.: MQTT-PWN, IoT exploitation & recon framework. https://mqtt-pwn.readthedocs.io/en/latest/index.html (2018). Accessed Feb 2020
3. Ahmed, N., De, D., Hussain, I.: Internet of things (IoT) for smart precision agriculture and farming in rural areas. IEEE Internet of Things J. **5**(6), 4890–4899 (2018)
4. Alansari, Z., Soomro, S., Belgaum, M.R., Shamshirband, S.: The rise of internet of things (IoT) in big healthcare data: review and open research issues. In: Saeed, K., Chaki, N., Pati, B., Bakshi, S., Mohapatra, D.P. (eds.) Progress in Advanced Computing and Intelligent Engineering, pp. 675–685. Springer, Singapore (2018). https://doi.org/10.1007/978-981-10-6875-1_66
5. Arasteh, H., Hosseinnezhad, V., Loia, V., Tommasetti, A., Troisi, O., Shafie-khah, M., Siano, P.: Iot-based smart cities: a survey. In: 2016 IEEE 16th International Conference on Environment and Electrical Engineering (EEEIC), pp. 1–6. IEEE (2016)

6. Barber, D.: Bayesian Reasoning and Machine Learning. Cambridge University Press, Cambridge (2012)
7. Dinculeană, D., Cheng, X.: Vulnerabilities and limitations of MQTT protocol used between IoT devices. Appl. Sci. **9**(5), 848 (2019)
8. Harsha, M.S., Bhavani, B.M., Kundhavai, K.R.: Analysis of vulnerabilities in MQTT security using shodan API and implementation of its countermeasures via authentication and ACLs. In: 2018 International Conference on Advances in Computing, Communications and Informatics (ICACCI), pp. 2244–2250 (2018)
9. Hindy, H., Brosset, D., Bayne, E., Seeam, A.K., Tachtatzis, C., Atkinson, R., Bellekens, X.: A taxonomy of network threats and the effect of current datasets on intrusion detection systems. IEEE Access **8**, 104650–104675 (2020)
10. Hindy, H., Hodo, E., Bayne, E., Seeam, A., Atkinson, R., Bellekens, X.: A taxonomy of malicious traffic for intrusion detection systems. In: 2018 International Conference On Cyber Situational Awareness, Data Analytics And Assessment (Cyber SA), pp. 1–4 (2018)
11. Hindy, H., Brosset, D., Bayne, E., Seeam, A., Bellekens, X.: Improving SIEM for critical SCADA water infrastructures using machine learning. In: Katsikas, S.K., Cuppens, F., Cuppens, N., Lambrinoudakis, C., Antón, A., Gritzalis, S., Mylopoulos, J., Kalloniatis, C. (eds.) Computer Security, pp. 3–19. Springer International Publishing, Cham (2019)
12. Hindy, H., Tachtatzis, C., Atkinson, R., Bayne, E., Bellekens, X.: MQTT-IOT-IDS2020: MQTT internet of things intrusion detection dataset. IEEE Dataport (2020). https://doi.org/10.21227/bhxy-ep04
13. Hodo, E., Bellekens, X., Hamilton, A., Dubouilh, P.L., Iorkyase, E., Tachtatzis, C., Atkinson, R.: Threat analysis of IoT networks using artificial neural network intrusion detection system. In: 2016 International Symposium on Networks, Computers and Communications (ISNCC), pp. 1–6. IEEE (2016)
14. Hosmer, D.W., Lemeshow, S., Sturdivant, R.X.: Applied Logistic Regression, vol. 398. Wiley, Hoboken (2013)
15. Larose, D.T., Larose, C.D.: Discovering Knowledge in Data: An Introduction to Data Mining. Wiley, Hoboken (2014)
16. Lior, R.: Data Mining with Decision Trees: Theory and Applications, vol. 81. World scientific, Toh Tuck (2014)
17. Nogues, M., Brosset, D., Hindy, H., Bellekens, X., Kermarrec, Y.: Labelled network capture generation for anomaly detection. In: International Symposium on Foundations and Practice of Security, pp. 98–113. Springer (2019)
18. Ring, M., Wunderlich, S., Grüdl, D., Landes, D., Hotho, A.: A toolset for intrusion and insider threat detection. In: Palomares, C.I., Kalutarage, H., Huang, Y. (eds.) Data Analytics and Decision Support for Cybersecurity Data Analytics, pp. 3–31. Springer International Publishing, Cham (2017). https://doi.org/10.1007/978-3-319-59439-2_1
19. Stanford-Clark, A., Truong, H.L.: Mqtt for sensor networks (MQTT-SN) protocol specification. International business machines (IBM) Corporation version **1**, 2 (2013)

20. Steinwart, I., Christmann, A.: Support Vector Machines. Springer Science & Business Media (2008)
21. VanderPlas, J.: Python Data Science Handbook: Essential Tools for Working with Data. O'Reilly Media Inc, Sebastopol (2016)

Smart Lamp or Security Camera? Automatic Identification of IoT Devices

Mathias Dahl Thomsen[1], Alberto Giaretta[2(✉)], and Nicola Dragoni[1,2]

[1] DTU Compute, Technical University of Denmark, Kongens Lyngby, Denmark
mdahlthomsen@hotmail.com, ndra@dtu.dk

[2] Centre for Applied Autonomous Sensor Systems (AASS), Örebro University, Örebro, Sweden
alberto.giaretta@oru.se

Abstract. The tsunami of connectivity brought by the Internet of Things is rapidly revolutionising several sectors, ranging from industry and manufacturing, to home automation, healthcare and many more. When it comes to enforce security within an IoT network such as a smart home, there is a need to automatically recognise the type of each joining devices, in order to apply the right security policy. In this paper, we propose a method for identifying IoT devices' types based on natural language processing (NLP), text classification, and web search engines. We implement a proof of concept and we test it against 33 different IoT devices. With a success rate of 88.9% for BACnet and 87.5% for MUD devices, our experiments show that we can efficiently and effectively identify different IoT devices.

Keywords: Internet of Things · Device identification · Profiling · Natural language processing · Text classification

1 Introduction

With the increased popularity of the Internet of Things (IoT) and a rapidly expanding market, there is a growing demand for new IoT devices. On the one hand, this creates an opportunity for companies and start-ups to cover such demand, capitalise on this trend, and improve cooperation. In 2017 alone, 2888 North American startups managed to attract investments for $125 billion. All combined, these 2888 startups grew up to $613 billion and 95 of them managed to grow up to $1 billion each. Last, investors play a fundamental role in transferring knowledge to and among startups [14].

On the other hand, the massive market demand pressures manufacturers to provide better features than competitors, quicker, and for lower prices. Unfortunately, this led many manufacturers to overlook the security of their devices, while focusing on more apparent features. As a result, many IoT devices currently operating are insecure, easy to penetrate, and to exploit for malicious activities [7]. For example, botnet malwares recruit devices for striking Distributed Denial of Service (DDoS) attacks. Using this technique, in 2016 the

B. Ghita and S. Shiaeles (Eds.): INC 2020, LNNS 180, pp. 85–99, 2021.
https://doi.org/10.1007/978-3-030-64758-2_7

Mirai malware took advantage of unprotected IoT devices and performed the largest DDoS attack ever recorded [5].

IoT devices specialise on a small set of tasks, collaborating with other devices for providing complex services (e.g., a set of sensors in a smart home). One critical point is that current devices do not specify which entities they collaborate with, in order to function properly [9]. This results in substrata of hidden behaviours and communications, hard to manage and reconcile with networks' security requirements. Expressing behaviours is a crucial aspect of identity management. As pointed out by Roman et al. [19], identity management is not only a problem of binding unique IDs with devices, but also of describing the core features of such devices.

Protocols and standards, such as Building Automation and Control networks (BACnet) [11] and Manufacturer Usage Descriptions (MUD) [13], partially address this issue. For instance, MUD devices explicitly describe their requirements through a MUD file. A network must be equipped with a node (the MUD Controller) which verifies MUD files and stores the network Access Control List (ACL). MUD is limited to domains, ports, and protocols used by IoT devices for their communications. This makes hard for network admins to reason about categories of devices, and to setup concise and intelligible network policies.

Following the line of explicit behavioural description, Giaretta et al. [9,10] have recently proposed to apply the Security-by-Contract paradigm (S×C) to IoT. In a nutshell, IoT devices store a contract which describes their security-relevant behaviour, easy to check and verify against network security policies stored on a Fog node. In the past, S×C paradigm has been successfully applied to different fields, ranging from pervasive mobile computing [6], through web applications, to multi-application smart cards.

All these efforts share a common drawback: what happens if a device does not carry the behavioural description required by the network it is joining? For example, an IoT device might be MUD-compliant, but not S×C-compliant. In the name of compatibility, it is desirable to allow devices to join the network with a minimal set of permissions, according to the nature of the devices themselves. This raises the problem of identifying devices' type. How can a network node, such as a Fog node controlling a smart home, discern an indoor lamp from a smart lock?

Contribution. In this paper, we propose an approach for identifying IoT devices, and we apply it to MUD and BACnet devices. In particular, we describe a technique that identifies devices' types by means of text classification algorithms and natural language processing (NLP). First, we perform experiments in order to identify which combination of algorithms identify devices best. Then, we evaluate our proof of concept (PoC), focusing on accuracy and speed. We also analyse the downsides of our implementation, and how to amend them.

Outline of the Paper. In Sect.2 we motivate our work. In Sect. 3 we detail the necessary steps to identify an IoT device. In Sect. 4 we discuss a first implementation of our PoC, and in Sect. 5 we evaluate its performance. In Sect. 6 we highlight the potential legal issues of our system and how they can be amended. Also, we show that the second PoC shows better detection rates than the first version. In Sect. 7 we discuss benefits and shortcomings of our work and in Sect. 8 we give a brief overview of related work. Last, we wrap up our contribution in Sect. 9.

2 Motivation

Let us assume that we add an IoT indoor lamp to our local network. It is reasonable to suppose that a lamp interacts with motion sensors, so that the light can be turned on and off, depending on our presence. It is equally reasonable to suppose that an indoor lamp does not interact with external security cameras.

The entire reasoning is based on the nature of the device, a smart lamp. It is irrelevant that the lamp is manufactured by *Philips* or *Samsung*, or that it is *Model A7* or *Model T1*. There are cases where specific policies are necessary for specific devices, while other cases require a reasoning based on the task the devices perform (e.g., lighting up a room). Most protocols and paradigms, such as BACnet and MUD, do not enforce any information regarding the nature of the device.

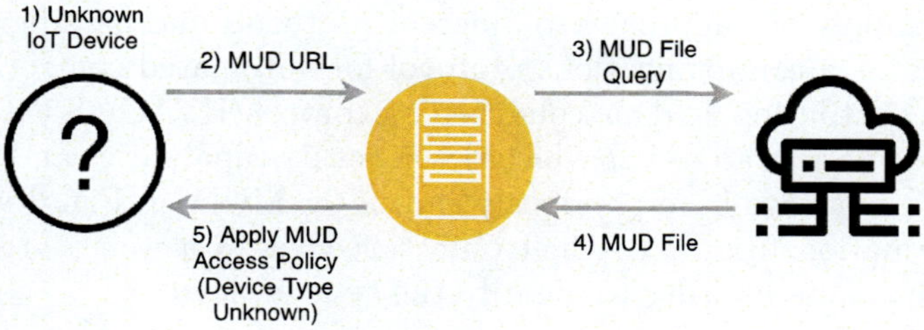

Fig. 1. Basic MUD architecture.

This is clearer by looking at the basic MUD architecture shown in Fig. 1. In short, the MUD device provides a MUD URL to the MUD Controller. In turn, the controller queries the URL, downloads the MUD file, and verifies it. If everything is correct, it creates a new access policy for the IoT device and then allows it to the network. The device evaluation is based on the fact that the manufacturer defined and signed a MUD file. No step of the MUD paradigm requires to know which device it deals with. This makes hard for network administrators to define security policies at the category level. How can we enforce a *lamp-policy* on our smart lamp, if our system cannot tell a lamp from a camera?

In Fig. 2 we show an example of how device identification could work with MUD and help administrators to apply policies according to categories. Once that MUD file has been retrieved, the Fog Node extracts and organises the relevant information. The Fog Node then feeds the data to a web search engine, gathers more information about the device, and combines everything together.

The identification phase takes place and the network is now able to enforce security policies predefined for the appropriate category. If no policies have been defined for the category, or the identification attempt fails, the Fog Node can still apply the MUD file as-is.

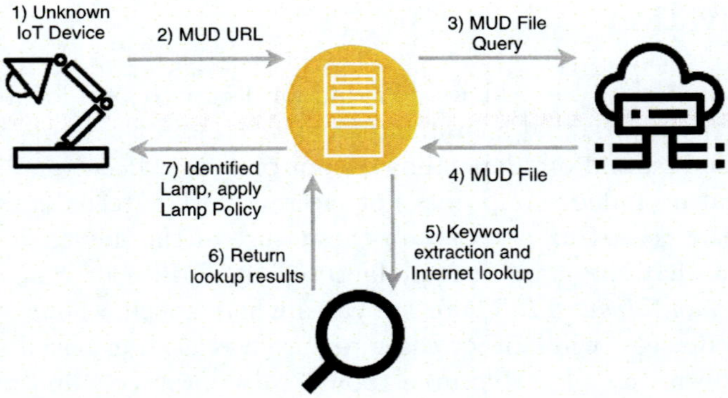

Fig. 2. MUD architecture integrated with our proposal.

We can apply our approach to different protocols and paradigms. For example, BACnet is a communications protocol for automated control systems, such as heating, ventilating, and air-conditioning control (HVAC). Each BACnet compliant device carries an *object* which describes its capabilities and relevant data. Similarly to what we have previously shown for MUD, in Fig. 3 we extract relevant information from a BACnet object, feed it to a web search engine, and process the results in order to identify the type of device.

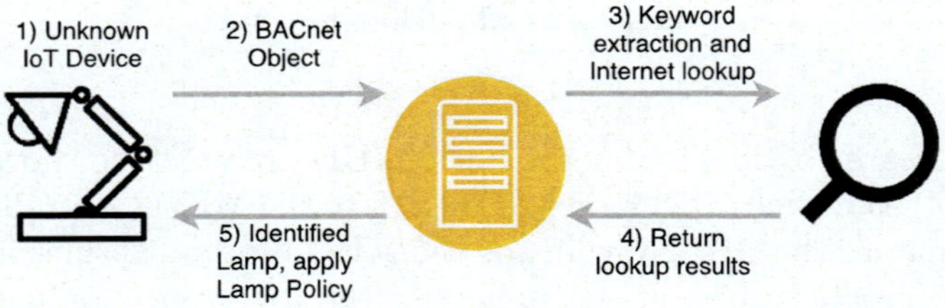

Fig. 3. BACnet integrated with our proposal.

3 Device Classification

Network traffic analysis is not suitable for quickly identifying IoT devices. Capturing and analysing pcap traces can require many hours of unrestricted interactions. This is not desirable, if such classification is necessary for enforcing network security policies, when a device joins the network.

In this paper, we propose a device classification approach which is composed of two main parts: the information retrieval and the text classification. As shown in Fig. 4, the first part (phase 1 and phase 2, in red) is responsible for extracting relevant texts from the device information fields. This part heavily depends on the IoT protocol stack, therefore it cannot be generalised for every IoT device. Phase 3, coloured in green in Fig. 4, is responsible for extracting relevant keywords from the texts (phase 3A), feeding them to web search engines and extracting the search results (phase 3B), and finally classifying the device (phase 3C).

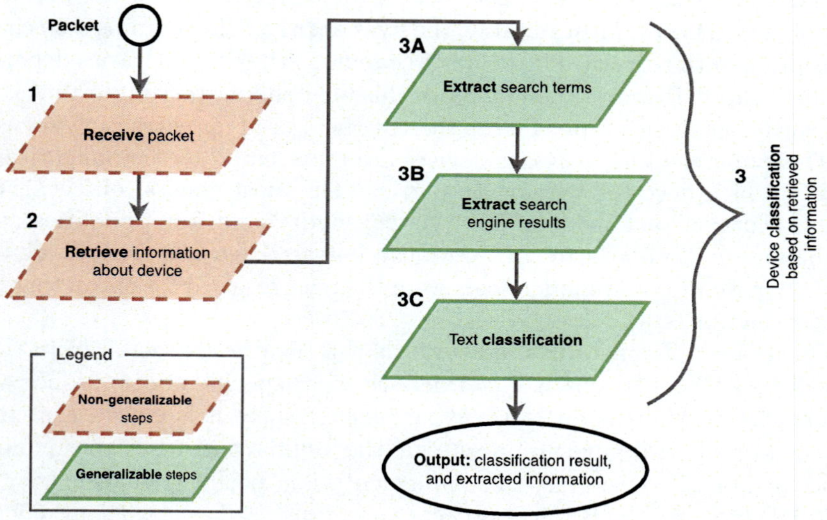

Fig. 4. Overview of a device profiling steps. The red steps cannot be generalised, since they depend on the specific protocol stack. The green steps, all part of the text classification phase, are generalisable.

In our vision, our implementation would run on a Fog node within a local network. Due to their computing power and their capability of offloading tasks to the Cloud layer [15,18], Fog nodes are central elements for different frameworks, such as S×C [9]. In particular, our solution would rely on the upper Cloud layer (we refer the reader to the architectural model in [15]) to gather relevant information and extract the device type.

Even though it is out of the scope of this work, we highlight that this technique could fuel other interesting applications within other frameworks (such as S×C). For example, the gathered data could provide some information about devices behaviour and enable the framework to create tailored contracts. Else, the framework could store predefined contracts for each IoT device type (e.g., lamp, camera, smart plug, etc.), and assign the correct one to the profiled device, based on the classification result. In any case, it is fundamental to correctly identify the device type.

4 Proof of Concept

In this Section, we describe the PoC we developed in Python for assessing our approach. The source code and the data sets are publicly available on GitHub [1].

In Fig. 4 we noted that the initial steps cannot be generalised. With regard to the information retrieval step, we have to narrow down our scope and choose relevant standards and protocols. For general consumer IoT devices we focus on the Manufacturer Usage Description (MUD) standard. For industrial IoT devices, we take into consideration the BACnet protocol. The overall approach for extracting information is therefore different.

MUD files contain various fields that provide relevant information for detecting the device types. Most notably, the `systeminfo` field contains a 60 characters description of the device. The field is free text, so the content varies depending on the file creator. However, we can parse this field and extract important keywords to identify the device type. This is the only field used for extracting keywords for MUD capable devices. BACnet devices are represented by virtual entities called *objects*, which describe various properties. The most general object, the *device object*, contains text fields about vendor, model, and functionalities exhibited by the device. Similarly to MUD, we can extract keywords from these texts. In particular, vendor and model fields were the ones we used for extracting BACnet devices keywords.

There are differences in the keywords that can be extracted from the MUD files and BACnet objects. It is not defined what specific information is specified in *systeminfo*, whereas for the BACnet *device object* it is known that the fields will contain specific information about the vendor and what the model of the device is. For the purpose of this implementation, both the fields extracted from the MUD files and the BACnet objects are simply used as is without any attempt to discard irrelevant text from the field.

To gather as much information as possible, we feed the extracted keywords to Google and Bing. Our goal is to increase the amount of information at our disposal, so that we improve our chances of identifying the devices. Obtained all the relevant text, we can start the classification phase.

4.1 Classification of Device Type

In our work, we have considered various text classification models and algorithms for identifying the device type. However, classification models cannot understand free text. We need to convert the text to feature vectors. This step is called feature extraction. As we show in Table 1, we evaluated different NLP approaches, such as Term Frequency (TF) and Term Frequency-Inverse Document Frequency (TF-IDF). We also evaluated various configurations of N-grams and custom Lemmatizers.

Our tests were based on a Support Vector Machine (SVM) classification model, used as a baseline for each test. The dataset used for training and testing

[1] Publicly released at: https://github.com/MDThomsen/DBAC-Device-Detection.

Table 1. Tested NLP approaches for feature extraction

Test	Precision Avg	Recall Avg	F_1 Score Avg	Support
TF	0.88	0.75	0.76	11.7
TF + Bigram	0.86	0.75	0.75	11.1
TF + Lemmatizer	0.80	0.71	0.70	10.2
TF-IDF	0.91	0.83	0.85	11.5
TF-IDF + Bigram	0.89	0.82	0.83	11.3
TF-IDF + Lemmatizer	0.90	0.82	0.83	11.3
doc2vec	0.78	0.40	0.47	20.0

the feature extraction methods was based on approximately 100 different texts, manually created. Each text was assigned to 1 out of 10 different categories of general consumer IoT devices. We performed a k-fold test (where k = 8), and for each fold we calculated precision, recall, and F_1 score. Therefore, F_1 *Score Avg* is not calculated over *Precision Avg* and *Recall Avg*, but is the average of all F_1 scores, previously calculated. TF-IDF scored better than TF, with respect to precision, recall, and F_1 score. Using bigrams to extend the contextual information did not improve the accuracy of neither TF nor TF-IDF. Performing extra preprocessing of the text with a lemmatizer did not provide any gains in accuracy; this shows that additional modelling or preprocessing are unnecessary, for this specific instance. For implementing our PoC, we opted for TF-IDF.

Table 2. Tested text classifiers

Classifier Testing	Precision Avg	Recall Avg	F_1 Score Avg	Support
LibLinear Logistic Regression	0.89	0.86	0.84	19.3
SGD Logistic Regression	0.93	0.88	0.88	20.7
SGD SVM	0.94	0.91	0.91	20.6
SVM (LibSVM)	0.95	0.92	0.92	20.9
MLP	0.86	0.76	0.76	29.9
Random Forest	0.92	0.88	0.88	19.6
XGBoost Random Forest	0.89	0.87	0.85	19.7

Similar to what we have done for feature extraction approaches, we compared different text classification algorithms. In particular, we considered Logistic Regression, Support Vector Machines, Neural Networks, and Random Forests. For Logistic Regression, we considered 2 models. One based on the LibLinear implementation [8], the other one based on stochastic gradient descent (sgd) [17]. For Support Vector Machines, we based the first model on sgd [17], and the second one on an SVM implementation called LibSVM [3]. The neural network we used for testing was a multi-layer perceptron (MLP) classifier [17] with one hidden layer with 100 perceptrons in the hidden layer. Last, for Random Forests we tested two different implementations. One which averages the results of every tree [17], and another one based on gradient boosting [4].

As shown in Table 2, most classifiers performed well. Support vector machines shine in cases where the feature vectors have high dimensionality, which is the case for text classification [17]. Logistic Regression models share many advantages with SVMs, and performed well enough. With regard to random forests, even though they both performed well, they need to be configured properly to avoid overfitting. The MLP model performed the worst, likely due to the small dataset we used for training the model. With a larger training dataset, we believe that the results would have been better. Given the current results and the theoretical advantages brought by SVMs, we opted for using the LibSVM model.

5 Evaluation

After we have chosen the appropriate feature extraction and text classification algorithms, we defined a set of tests. Our goal is to verify that our approach can identify correctly MUD and BACnet devices. First, we created 2 different training datasets in order to train the classification models. One dataset contains texts for 10 categories of general consumer IoT devices. The other dataset contains these 10 categories of texts, plus 2 categories specific for industrial IoT devices. The final categories are:

- **General devices** Alarm, Camera, Doorbell, Electric Control, Health Monitor, House Environment Monitor, Light, Media Player, Motion Sensor, Speaker.
- **Industrial devices** Industry Environment Sensor, Industry Controller.

Additionally, we have created 2 testing datasets. The first dataset contains 24 general consumer IoT devices, in the form of MUD files generated by Hamza et al. [12]. The second dataset contains the first dataset items, plus 9 industry IoT devices in the form of BACnet objects. This led to the following tests:

- **MUD-:** Testing MUD devices. Training dataset contains only texts and categories of general IoT devices.
- **MUD+:** Testing MUD devices. Training dataset contains texts and categories of both general and industry IoT devices.
- **BACnet+:** Testing BACnet devices. Training dataset contains texts and categories of both general and industry IoT devices.

The MUD- test evaluates how well the implementation performs when we only consider MUD devices and categories of devices targeted at general consumers. As an extension, with MUD+ we evaluated the same set of IoT devices, but we expanded the categories to include industrial IoT devices.

We did not perform BACnet- tests: with only two categories for industrial IoT devices, the training dataset would have been too small. Last, in BACnet+

we tried to identify BACnet-enabled devices using the same categories we used for MUD+. In this way, we were also able to test the generalisability of our approach.

In the first implementation phase, we extract relevant keywords from MUD and BACnet devices, we automatically feed them to search engines (i.e., Google and Bing), and we collect the resulting texts. Then, thanks to the classification model, we classify each text with a score between 0.0 and 1.0, and we choose the highest scoring one for classifying the device. We have also defined a minimum threshold for discarding every irrelevant text. In case that all the texts score below this threshold, we consider the device non-classified. Initially this threshold was set to 0.1.

Initial tests gave an accuracy of 75% for MUD+, 83% for MUD-, and 33% for BACnet+. The small dataset we used, as well as the similarities between some categories, negatively affected the result. In particular, some device types were similar in terms of features and some categories shared the same vocabulary. Due to these similarities, the classification scoring would often fail. To alleviate these issues, first we increased the dataset size. Also, we further manually cleaned the texts, in order to remove irrelevant terms.

Some problems also derived from basing the classification on the highest scoring text. Therefore, we opted for classifying each text and then accumulating the score towards each category. To increase separation between higher and lower scoring texts, each classification score is squared. We named this approach Cumulative Squared Scoring (CSS). The CSS approach achieved higher accuracy for both MUD and BACnet, as shown in Table 3. For both MUD tests one device classification was discarded (marked as "No Classification"). This was due to similarities between categories, leading to a classification falling below our pre-defined threshold.

6 Legal Issues

In our first PoC, we scraped the URLs retrieved from the search engines. This was a problem, since scraping often breaches the websites' terms of service (TOS). Many lawsuits regarding scraping have been filed, with varying outcomes [2]. There are several cases in the United States, where businesses scraped competitors' websites for financial gain. Such lawsuits ended in favour of the scraped

Table 3. Results obtained with CSS and a cleaned data set

CSS	MUD-	MUD+	BACnet+
Accuracy	0.88	0.88	0.89
No Classification	0.04	0.04	0.00
Average Classification Score	1.17	1.09	1.22

party. Other cases focused on whether or not freely public data should be scrapable [2]. Overall, there seems to be huge variability about the outcomes, depending on specific laws for each country.

In order to minimise any potential legal issue, we modified our implementation. Both Bing and Google return small snippets of text that summarise the contents of the indexed web pages. We decided to use the snippets directly, sidestepping the scraping phase. Our initial dataset were also created through scraping. We replaced all the problematic texts with other texts, scraped from more permissive websites. After testing various thresholds for discarding texts, a new threshold of 0.2 was used, rather than 0.1. For the sake of clarity, we called this approach TOS. Snippets not only protected us from TOS issues, but also improved our results. As we can see in Table 4, the accuracy improved, compared to Table 3, and the average classification score is higher for each and every item.

Table 4. Results obtained with search engines snippets, cumulative scoring, and a cleaned data set, denoted TOS

TOS	MUD-	MUD+	BACnet+
Accuracy	0.96	0.88	0.89
No Classification	0.00	0.04	0.00
Average Classification Score	2.53	1.82	1.63

Moreover, by using snippets and avoiding the scraping phase, the average run time was dramatically reduced. As shown in Table 5, the average run time for identifying MUD devices is 10× shorter with snippets. Similarly, the average run time for BACnet devices is 7× shorter. In terms of devices correctly identified, our final TOS solution allowed us to correctly identify almost 90% of our MUD and BACnet devices, as shown in Table 6. In Table 7, we provide the detailed list of which devices we were able to correctly identify, and which ones not. In the third column we specified the actual type of device, while in the fourth column we report the output of our classification proof-of-concept. While three of the four failed attempts were plain wrong classifications, one of the devices (i.e., the *Blipcare BP Meter*) were not classified at all. As aforementioned, some categories are considerably similar to each other; in this particular case, the device scored equally in different categories and failed to meet the minimum threshold we set.

With regard to increased accuracy and classification score shown in Table 4, we give credit to the relevancy and conciseness of the snippets returned from the search engines. Both Bing and Google have methods for summarising the websites through machine learning algorithms, trained to extract the most relevant words from long texts. The results we obtained indicate that the snippets accurately portray the websites' contents, providing relevant text and very little

[2] https://www.tripwire.com/state-of-security/featured/hiq-v-linkedin-controls-publicly-available-data/.

Table 5. Average time required for classifying devices with CSS and TOS

Device	CSS (s)	TOS (s)
MUD	20.24	2.78
BACnet	48.30	7.62

Table 6. Results obtained with TOS technique

Protocol	Devices identified (%)
MUD	87.5
BACnet	88.9
Total	87.9

noise. This also shows that more text does not entail better results. However, if a short text (such as a snippet) contains noise, the classification is affected to a higher degree than if it were dealing with a longer text.

7 Discussion

While our implementation performs well, there is room for improvement. Typically, training data sets contain several thousands of texts. However, in our experiment we had to manually create our training dataset, resulting in 5000 to 10000 words per category. This shortage impacted the choice of the classification model, as a larger dataset might have lead to different test results. Moreover, the testing dataset does not have a consistent amount of devices across classes. The test set has 24 MUD devices but only 9 BACnet devices, making some categories not well represented.

There is also variability in performance, depending on the MUD files and the BACnet objects. As we discussed before, we extract the first set of keywords from specific fields. In the case of MUD, the *systeminfo* field is optional (refer to the specification[3]) so it might not be possible to identify the device with our approach. The same goes for BACnet: the descriptive fields in the *device object* do not have specific requirements, which could lead to potential issues.

Furthermore, our solution is tied to the correctness of the search engines results. If search engines provide text that is irrelevant or misleading, the classification of the device type might be affected. Let us consider the case of *Google bombing*, the act of boosting a website pagerank for unrelated search terms, by massively linking from external webpages [1]. As a main side effect of this practice, search results are polluted with a considerable amount of noise. If we try to classify a security camera, and an adversary manipulated the results such that websites describing lamps (instead of cameras) are showed, our algorithm might lead to a misclassification. In turn, the misclassification might lead to security issues.

[3] https://datatracker.ietf.org/doc/rfc8520/.

Table 7. Final results obtained with our identification technique

Protocol	Device name	Type	Predicted Type	Result
MUD	Amazon Echo	Speaker	Speaker	Pass ✓
MUD	August Doorbell	Doorbell	Doorbell	Pass ✓
MUD	Awair Air Quality	House Environment Monitor	House Environment Monitor	Pass ✓
MUD	Belkin Camera	Camera	Camera	Pass ✓
MUD	Blipcare BP Meter	Health Monitor	No Classification	Fail ✗
MUD	Canary Camera	Camera	Camera	Pass ✓
MUD	Chromecast Ultra	Speaker	Speaker	Pass ✓
MUD	Dropcam	Camera	House Environment Monitor	Fail ✗
MUD	Phillips Hue Bulb	Light	Light	Pass ✓
MUD	iHome Power Plug	Electric Control	Electric Control	Pass ✓
MUD	Lifx Bulb	Light	Light	Pass ✓
MUD	Nest Smoke Sensor	House Environment Monitor	House Environment Monitor	Pass ✓
MUD	Netatmo Camera	Camera	Camera	Pass ✓
MUD	Netatmo Weather	House Environment Monitor	House Environment Monitor	Pass ✓
MUD	Ring Doorbell	Doorbell	Doorbell	Pass ✓
MUD	Samsung Smart Cam	Camera	Camera	Pass ✓
MUD	TP Link Camera	Camera	Camera	Pass ✓
MUD	TP Link Plug	Electric Control	Electric Control	Pass ✓
MUD	Triby Speaker	Speaker	Speaker	Pass ✓
MUD	Wemo Motion	Motion Sensor	Motion Sensor	Pass ✓
MUD	Wemo Switch	Electric Control	Electric Control	Pass ✓
MUD	Withings Baby Monitor	House Environment Monitor	Health Monitor	Fail ✗
MUD	Withings Cardio	Health Monitor	Health Monitor	Pass ✓
MUD	Withings Sleep Sensor	Health Monitor	Health Monitor	Pass ✓
BACnet	AROB Universal Room Controller	Industry Controller	Industry Controller	Pass ✓
BACnet	BACnet Dewpoint Transmitter	Industry Environment Sensor	Industry Environment Sensor	Pass ✓
BACnet	BAC-RI Room Interface Module	Industry Controller	Industry Controller	Pass ✓
BACnet	BACnet Room Pressure Monitor	Industry Environment Sensor	Health Monitor	Fail ✗
BACnet	SRI-70 Analogue Room Interfaces	Industry Controller	Industry Controller	Pass ✓
BACnet	CDD3 CO2 Detector	Industry Environment Sensor	Industry Environment Sensor	Pass ✓
BACnet	CDR-BAC Room CO2/Temp Sensor	Industry Environment Sensor	Industry Environment Sensor	Pass ✓
BACnet	SRC-100 Series Zone Controllers	Industry Controller	Industry Controller	Pass ✓
BACnet	Touchplate Light Controller	Industry Controller	Industry Controller	Pass ✓

8 Related Work

In this paper, our goal is to infer the type of IoT devices. Manufacturer Usage Descriptions (MUD) [13] is an IETF specification (RFC8520), in which manufacturers specify which hosts and ports their devices need for operating correctly.

Even though MUD does not allow exhaustive behavioural descriptions, MUD files can be used for profiling IoT devices without time-consuming phases for collecting devices' traffic packets.

Hamza et al. [12] proved that is possible to convert pcap traces to behavioural description files, such as MUD. The authors describe their conversion tool and showcase its use by creating MUD profiles for 28 popular IoT devices. These devices are a good base for testing our implementation. Other works show that is possible to detect a device type from its network traffic [20]. This is done by monitoring different variables, such as the amount of data, communication patterns, idle time, port numbers, DNS queries, and so on. By collecting months of data, the authors were able to identify various devices through their network traffic.

From a larger perspective, researchers proposed different approaches for achieving reliable fingerprints of IoT devices. Yang et al. [21] proposed a triplet label (type,vendor,product) for identifying IoT devices, and a fingerprint generation approach based on neural networks. Meidan et al. [16] proposed ProfilIoT, a tool based on supervised learning, capable of identifying accurately IoT devices. However, it is unclear if such approaches would be feasible for achieving near real-time fingerprinting, which is a critical characteristic for managing IoT devices joining a network for the first time.

9 Conclusion

In this paper, we have proposed a technique for device identification which combines natural language processing, text classification, and web search engines. First, we tested different feature extraction and text classification models, opting for combining TF-IDF and Support Vector Machines. Second, we identified two web search engines and, in order to avoid terms of services issues, we opted for gathering data from the result snippets instead of scraping websites.

Then, we implemented a PoC and we applied it to various IoT devices. With a success rate of 87.5% for MUD devices, and 88.9% for BACnet devices, our experiments showed that not only it is possible to identify devices' types, but it can also be done in a matter of a few seconds. In future works, we plan to extend our PoC to Shodan (a web search engine dedicated to IoT devices), and to assess the robustness of our approach when evaluating non-IoT devices.

References

1. Adali, S., Liu, T., Magdon-Ismail, M.: An analysis of optimal link bombs. Theor. Comput. Sci. **437**, 1–20 (2012)
2. Broucke, S.V., Baesens, B.: Practical Web Scraping for Data Science: Best Practices and Examples with Python. Apress, New York, NY (2018)

3. Chang, C.C., Lin, C.J.: LIBSVM: a library for support vector machines. ACM Trans. Intell. Syst. Technol. **2**(3), 1–27 (2011)
4. Chen, T., Guestrin, C.: XGBoost: A scalable tree boosting system. In: Proceedings of the 22nd ACM SIGKDD International Conference on Knowledge Discovery and Data Mining. KDD 2016, ACM, New York, NY, USA (2016)
5. De Donno, M., Dragoni, N., Giaretta, A., Spognardi, A.: DDoS-capable iot malwares: comparative analysis and Mirai investigation. Secur. Commun. Netw. **1–30**(02), 2018 (2018)
6. Dragoni, N., Massacci, F., Walter, T., Schaefer, C.: What the heck is this application doing? a security-by-contract architecture for pervasive services. Comput. Secur. **28**(7), 566–577 (2009)
7. Dragoni, N., Giaretta, A., Mazzara, M.: The internet of hackable things. In: Proceedings of 5th International Conference in Software Engineering for Defence Applications, pp. 129–140. Springer International Publishing, Cham (2018)
8. Fan, R.E., Chang, K.W., Hsieh, C.J., Wang, X.R., Lin, C.J.: LIBLINEAR: a library for large linear classification. J. Mach. Learn. Res. **9**, 1871–1874 (2008)
9. Giaretta, A., Dragoni, N., Massacci, F.: IoT security configurability with security-by-contract. Sensors **19**(19), 4121 (2019)
10. Giaretta, A., Dragoni, N., Massacci, F.: Protecting the internet of things with security-by-contract and fog computing. In: 2019 IEEE 5th World Forum on Internet of Things (WF-IoT), pp. 1–6. IEEE, New York, NY (2019)
11. Haakenstad, L.K.: The open protocol standard for computerized building systems: BACnet. In: Proceedings of the 1999 IEEE International Conference on Control Applications, vol. 2, pp. 1585–1590. IEEE, New York, NY, August 1999
12. Hamza, A., Ranathunga, D., Gharakheili, H.H., Roughan, M., Sivaraman, V.: Clear As MUD: generating, validating and applying IoT behavioral profiles. In: Proceedings of the 2018 Workshop on IoT Security and Privacy. IoT S&P 2018, ACM, New York, NY, USA (2018)
13. Lear, E., Weis, B.: Slinging MUD: manufacturer usage descriptions: how the network can protect things. In: International Conference on Selected Topics in Mobile Wireless Networking (MoWNeT), pp. 1–6. IEEE, New York, NY (2016)
14. Lim, S., Kwon, O., Lee, D.H.: Technology convergence in the internet of things (IoT) startup ecosystem: a network analysis. Telematics Inform. **35**(7), 1887–1899 (2018)
15. Mahmood, Z.: Fog Computing: Concepts Frameworks and Technologies. Springer, Cham (2018)
16. Meidan, Y., Bohadana, M., Shabtai, A., Guarnizo, J.D., Ochoa, M., Tippenhauer, N.O., Elovici, Y.: Profiliot: A machine learning approach for iot device identification based on network traffic analysis. In: Proceedings of the Symposium on Applied Computing, pp. 506–509. SAC 2017, Association for Computing Machinery, New York, NY, USA (2017)
17. Pedregosa, F., Varoquaux, G., Gramfort, A., Michel, V., Thirion, B., Grisel, O., Blondel, M., Prettenhofer, P., Weiss, R., Dubourg, V., Vanderplas, J., Passos, A., Cournapeau, D., Brucher, M., Perrot, M., Duchesnay, E.: Scikit-learn: machine learning in Python. J. Mach. Learn. Res. **12**(85), 2825–2830 (2011)
18. Rayes, A., Salam, S.: Internet of Things From Hype to Reality: The Road to Digitization. Springer International Publishing, Cham (2019)
19. Román-Castro, R., López, J., Gritzalis, S.: Evolution and trends in IoT security. Computer **51**(7), 16–25 (2018)

20. Sivanathan, A., Gharakheili, H.H., Loi, F., Radford, A., Wijenayake, C., Vishwanath, A., Sivaraman, V.: Classifying IoT devices in smart environments using network traffic characteristics. IEEE Trans. Mobile Comput. **18**(8), 1745–1759 (2019)
21. Yang, K., Li, Q., Sun, L.: Towards automatic fingerprinting of iot devices in the cyberspace. Comput. Netw. **148**, 318–327 (2019)

Intelligent Self-reliant Cyber-Attacks Detection and Classification System for IoT Communication Using Deep Convolutional Neural Network

Qasem Abu Al-Haija(✉), Charles D. McCurry, and Saleh Zein-Sabatto

Department of Electrical and Computer Engineering, Tennessee State University, 3500 John A. Merritt Blvd, Nashville, TN 37209, USA
Qabualha@Tnstate.edu

Abstract. Internet of Things (IoT) is a promising profound technology with tremendous expansion and effect. However, IoT infrastructures are vulnerable to cyber-attacks due to the constraints in computation, storage, and communication capacity for the endpoint devices such as thermostat, home appliance, etc. It was reported that 99% of the cyber-attacks are developed by slightly mutating previously known attacks to generate a new attack tending to be handled as a benign traffic through the IoT network. In this research, we developed a new intelligent self-reliant system that can detect mutations of IoT cyber-attacks using deep convolutional neural network (CNN) leveraging the power of CUDA based Nvidia-Quad GPUs for parallel computation and processing. Specifically, the proposed system is composed of three subsystems: Feature Engineering subsystem, Feature Learning subsystem and Traffic classification subsystem. All subsystems are developed, verified, integrated, and validated in this research. To evaluate the developed system, we employed the NSL-KDD dataset which includes all the key attacks in the IoT computing. The simulation results showed a superior attacks' classification accuracy over the state-of-art machine learning based intrusion detection systems employing similar dataset. The obtained results showed more than 99.3% and 98.2% of attacks' classification accuracy for both binary-class classifier (normal vs anomaly) and multi-class classifier (five categories) respectively. All development steps and testing and verification results of the developed system are reported in the paper.

Keywords: Artificial intelligence · Convolutional neural network · IoT networks · Cyber-attack detection · Classification

1 Introduction

The Internet of Things (IoT) is comprised of a collection of heterogeneous resource-constrained objects interconnected via different network architectures such as wireless sensor networks (WSN) [1]. These objects or "things" are usually composed of sensors,

B. Ghita and S. Shiaeles (Eds.): INC 2020, LNNS 180, pp. 100–116, 2021.
https://doi.org/10.1007/978-3-030-64758-2_8

actuators, and processors with the ability to communicate with each other to achieve common goals/applications through unique "thing" identifiers with respect to the Internet Protocol (IP) [2, 3]. Current IoT applications include smart buildings, telecommunications, medical and pharmaceutical, aerospace and aviation, environmental phenomenon monitoring, agriculture, industrial processing and manufacturing and others. The basic IoT layered architecture is shown in Fig. 1. It has three layers: the perception layer (consist of edge-devices that interact with the environment to identify certain physical factors or other smart objects in the environment), the network layer (consists of a number of networking-devices that discover and connect devices over the IoT network to transmit and receive the sensed data), and the application layer (consists of various IoT applications/services that are responsible for data processing and storage). Indeed, most cyber-attacks target the application and network layers of the IoT system.

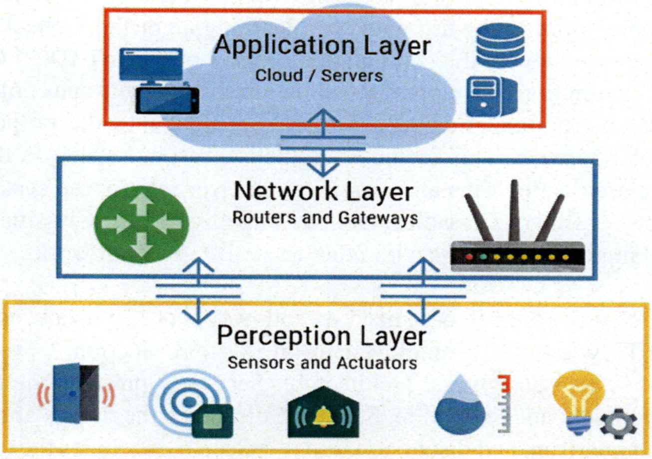

Fig. 1. IoT layered architecture [4].

IoT is a promising profound technology with tremendous expansion and effect. IoT infrastructures are vulnerable to cyber-attacks in that within the network simple endpoint devices (e.g. thermostat, home appliance, etc.) have constraints in computation, storage, and network capacity other than more complex endpoint devices, such as a smartphone, laptop etc., that may reside within the IoT infrastructure. Once the IoT infrastructure is breached, hackers have the ability to distribute the IoT data to unauthorized parties and can manipulate the accuracy and consistency of IoT data over its entire life cycle [5]. Therefore, such cyber-attacks need to be addressed for safe IoT utilization. Consequently, vast efforts to handle the security issues in the IoT model have been made in the recent years. Many of them were developed by coupling the field machine learning techniques with cybersecurity field. It should be noted that, the majority of IoT attacks are developed as slight deviations (i.e. mutations) of earlier known cyberattacks [6]. Such slight mutations of these IoT attacks have been demonstrated to be difficult to identify/classify using traditional machine learning techniques. Indeed, few promising state-of-art researches were conducted for cybersecurity using deep neural networks models such as [7–12].

In this paper, we propose a new intelligent system that can detect mutations of common IoT cyber-attacks using non-traditional machine learning techniques exploiting the power of Nvidia-Quad GPUs. The proposed system employs the convolutional neural network (CNN) along its associated machine learning algorithms to classify the NSL-KDD dataset records (we call our system as IoT-IDCS-CNN). The NSL-KDD dataset stores non-redundant records of all the key attacks of IoT computing with different levels of difficulties. Specifically, the main contributions of this paper can be summarized as follows:

- We provide a comprehensive efficient detection/classification model that can classify the IoT traffic records of NSL-KDD dataset into two (Binary-Classifier) or five (Multi-Classifier) classes. Also, we present detailed preprocessing operations for the collected dataset records prior to the use with deep learning algorithms. Besides, we illustrate a comprehensive view of the computation process of our IoT-IDCS-CNN.
- We provide a simplified development/validation environment and configurations along with an extensive simulation results to gain insight into the proposed model and the solution approach. This includes simulation results related to the classification accuracy, classification time and classification error rate for the system validation of both detection (Binary-Classifier) and classification (Multi-Classifier) in addition to benchmarking of our findings with other related state-of-art works.

The rest of this paper is organized as follows: Sect. 2 introduces and justify the dataset of IoT cyber-attacks employed in our system. Section 3 provides details of the proposed system design and architecture. Section 4 presents the environment for system development and validation. Section 5 discusses the details about experimental evaluation, comparison, and discussion. Finally, Sect. 6 concludes the findings of paper.

2 Dataset of Cyber-Attacks

NSL-KDD [13] is a reduced version of the original KDD'99 dataset [14] and consists of the same features as KDD'99. However, NSL-KDD includes more up-to-date and non-redundant attack records with different levels of difficulties. Figure 2 shows sample records of original KDD-NSL training dataset in CSV format but read by notepad in TXT format (prior to any processing technique). In this research, NSL-KDD dataset is collected and employed for many reasons including:

(a) It can be efficiently imported, read, preprocessed, encoded, and programed to produce two- or multi- class classification for IoT Cyber-attacks.
(b) It covers all key attacks of IoT computing including: Denial-of-Service (DoS) [15], Probe (side channel) [16], Root to Local (R2L) [17], User to Root (U2R) [17].
(c) It is obtainable as TXT/CSV filetype consisting a reasonable number of non-redundant records in the train and test sets. This improves the classification process by avoid the bias towards more frequent records.

```
0,tcp,ftp_data,SF,491,0,0,2,2,0.00,25,0.17,0.03,0.1
0,udp,other,SF,146,0,0,13,1,0.00,0,0.00,0.60,0.88,0
0,tcp,private,S0,0,0,0,123,6,1.00,26,0.10,0.05,0.00
0,tcp,http,SF,232,8153,0,5,5,0.20,55,1.00,0.00,0.03
0,tcp,http,SF,199,420,0,30,32,0.00,255,1.00,0.00,0.
0,tcp,private,REJ,0,0,0,121,19,0.05,19,0.07,0.07,0.
0,tcp,private,S0,0,0,0,166,9,1.00,9,0.04,0.05,0.00,
0,tcp,private,S0,0,0,0,117,16,1.00,15,0.06,0.07,0.0
0,tcp,remote_job,S0,0,0,0,270,23,1255,23,0.09,0.05,
```

Fig. 2. Sample records of KDD-NSL training dataset.

(d) It reveals high-level of IoT traffic structures and cyberattacks as well as it can be customized, expanded, and regenerated [13].

NSL-KDD dataset has been thoroughly developed with high-level diverse interpretations of the training data covering normal and abnormal IoT network traffic data. The normal data samples represent the legitimate data packets processed by the IoT network. The abnormal data samples represent mutated data packets (i.e., attacks) achieved by slight mutations in the previously developed attacks such as the small changes in the network packet header configurations. The original dataset is available in two classification forms: two-class traffic dataset with binary labels and multi-class traffic data set including attack-type labels and difficulty level. In both cases, it comprises 148,517 samples each with 43 attributes such as duration, protocol, service, and others [18]. The statistics of traffic distribution of NSL-KDD dataset is summarized in Table 1.

Table 1. Statistics of traffic distribution of NSL-KDD dataset [14].

	Two-classes dataset		Multi-classes dataset				
	Normal	Attack	Normal	DoS	Probe	R2L	U2R
Training	67343	58630	67343	45927	11656	995	52
Testing	9711	12833	9711	7458	2754	2421	200
Total	77,054	71,463	77,054	53,385	14,410	3,416	252

3 System Modeling

In this research, the proposed system is partitioned into distinct subsystems each of which is implemented with several modules. Specifically, the system is composed of three subsystems including: Feature Engineering (FE), Feature Learning (FL), and Detection and Classification (DC), as illustrated in Fig. 3.

Fig. 3. The three main subsystems composing the proposed system.

3.1 Implementation of Feature Engineering (FE) Subsystem

This subsystem is responsible for conversion of raw IoT traffic data records of NSL-KDD dataset into a matrix of labeled features that can be fed and trained by the neural network's part of the FL subsystem. The implementation stages of this subsystem include:

Importing NSL-KDD Dataset: In this stage, the collected dataset has been imported/read using MATLAB in a tabulated format instead of raw data in the original dataset text files. All data columns are assigned virtual names based on the nature of data in the cells. The imported dataset includes 43 different features/columns.

Renaming Categorial Features: Four of imported 43 features are categorial features that need to be renamed prior the data encoding and sample labeling processes. These features are the target protocol, the required service, the service flag, and the record category (e.g. normal or attack). Therefore, the four categorial columns have been renamed accordingly in this stage.

One Hot Encoding of Categorial Features: This module is responsible for conversion of the categorial data records into numerical data records in order to be employed by the neural network. Therefore, three categorial features undergo through One Hot Encoding process (1-N encoding) [19]. These features are the protocol column, the service column, and the flag column. The class feature/column is left for samples labeling process.

Labeling the Target Feature: This stage concerns with the samples labeling using numerical (integer) labels for the target classes. Therefore, the categorial 'Class Column' will be converted to numerical classes according to the classification technique. In our system implementation, we are considering two forms of traffic classifications: Binary classification (1: Normal vs. 2: Attack) and Multi classification (1: Normal, 2: DoS, 3:Probe, 4: R2L, 5: U2R). After this stage, all data records are available into numerical format (i.e. no categorial data are exist anymore). As a result of 1-N encoding and numerical labeling, we converted the dataset into 123 features and one data label.

Converting Tables to Double Matrix: At the end of dataset importing, encoding, and labeling processes, the dataset samples and targets should be provided to the neural network inputs of FL subsystem as a matrix of all input numerical samples. Therefore, encoded dataset tables have been converted to double matrix (148517 $\times$ 124).

Matrix Resizing with Padding Operation: This module is responsible to adjust the size of the dataset matrix to accommodate the input size for the FL subsystem. This was performed by resizing the matrix of the engineered dataset form 148517 × 124 to the new size of 148517 × 784, since the input size of every individual sample processed at FL subsystem is 28 × 28 (=784). Thereafter, the new empty records of this matrix were padded with zero-padding technique [20]. Besides, to avoid any feature biasing in the samples of the dataset, the padded records were distributed equally around the data samples.

Matrix Normalization with Min-Max Norm: Data normalization is performed to get all the data points to be in the same range (scale) with equal significance for each of them. Otherwise, one of the great value features might completely dominate the others in the dataset. Thus, this module is responsible to normalize all integer numbers of the dataset matrix into a range between 0–1 using Min-Max Normalization (MX-Norm) [21]. MX-Norm is well-known method to normalize data as its commonly used in machine learning applications. In this method, we scan all the all values in every feature, and then, the minimum value is converted into a 0 and the maximum value is converted into a 1, while the other values are converted (normalized) into a fraction value from 0 to 1.

Converting Tables to Double Matrix: This module is responsible to create the attack samples for the *CovNet* by reshaping the one-dimensional vectors of attack records into two-dimensional square matrices to accommodate the input size for the developed *CovNet* network. Accordingly, every one-dimensional vector sample (1 × 784) will be reshaped into two-dimensional matrix (28 × 28).

3.2 Implementation of Feature Learning (FL) Subsystem

So far, the FE subsystem has been developed and the next step is to process the encoded input features using FL subsystem-based CNN. The deep learning network will to be trained with minimum classification error and thus maximum accuracy. Generally, CNN involves of various layers including convolution, activation, pooling, flatten and others. Convolutional layers are the core constructing component of CNN network and they are hierarchically assembled to generate a number of feature-maps which enable CNNs to learn complex features being a vital operation to recognize patterns in the classification and detection tasks. Therefore, the FL subsystem is responsible to develop the appropriate *CNN* that can accept the encoded features from FE subsystem at the input layer and train on them with multiple hidden layers as well as updates the training parameters before classifying the IoT traffic dataset as normal or anomaly (mutated). The implementation stages of this subsystem include:

Feature Mapping with 2D-Convolution Operations Layer: This module is responsible to generate new matrices called feature maps that emphasizes the unique features of the original matrix [22]. These feature-maps are produced by convolving (multiply and accumulate) the original matrix ($n_{in} \times n_{in}$) using a number (N) of ($k \times k$) convolution filters with padding size (p) and stride size of (s) which yields the feature maps ($n_{out} \times n_{out}$). The size of the resultant feature maps can be evaluated as follows:

$$n_{out} = (n_{in} + 2p - k)/s + 1 \quad (1)$$

In this research, we have applied 20 convolution filters (9 × 9) over the 28 × 28 input samples with (p = 0, and s = 1) which resulted in 20 feature map each (20 × 20).

Feature Activation with ReLU Function: This module is responsible to activate all units of the feature maps with non-linear rectification function namely known as ReLU. ReLU function is MAX (X,0) that sets all negative values in the matrix X to zero while all other values are kept constant. The reason of using ReLU is that training a deep network with ReLU tended to converge much more quickly and reliably than training a deep network with other non-linear activation functions such sigmoid or tanh activation functions [23].

Down-Sampling with Pooling Operations Layer :This module is responsible to generate new matrices called pooled feature maps that reduces the spatial size of the rectified feature maps and thus reduces the number of parameters and computation complexity in the network [22]. This can be done by combining the neighboring points of a particular region of the matrix representation into a single value that represent the selected region. The adjacent points are typically selected from a fixed size-square matrix (determined according to the application). Among these points of the applied matrix, one value is nominated as the maximum or mean of the selected points. In this research, we have used the mean pooling technique to develop the pooling layer since it combines the contribution of neighboring points instead of only selecting the maximum point. To produce the pooled feature-maps ($L_{out} \times L_{out}$), the pooling filter ($f \times f$) is independently applied over the rectified feature-maps ($L_{in} \times L_{in}$) with stride (s) as follows:

$$L_{out} = (L_{in} - f)/s + 1 \quad (2)$$

In this research, we have applied 20 pooling operation (2 × 2) over the 20 × 20 rectified feature-maps with (s = 2) which resulted in 20 feature map each (10 × 10).

3.3 Implementation of Detection and Classification (DC) Subsystem

DC subsystem is responsible of providing traffic classification for the input traffic data into binary-class classification (2-Classes: normal vs. anomaly) or multi-class classification (5-Classes: Normal, DoS, Probe, R2L, U2R). This subsystem is composed of three consecutive stages as follows.

Flattening Layer of Pooled Feature Maps: This module is responsible to linearize the output dimension of the convolutional/pooling layers network to create a single long feature vector [22]. This can be achieved by converting the 2D data of N- Pooled feature-maps into a 1-D array (or vector) to be inputted to the next layer, which is connected to the final classification model, called a dense or fully connected layer. Since flatten layer collapses the spatial dimensions of the input into the channel dimension (array), this means that if the input to the flatten layer is (N) feature maps each with a dimension of $(F_{in} \times F_{in})$ then the flattened output (F_{out}) can be obtained by linear multiplication of the input dimensions by the number of maps, that's it:

$$F_{out} = N \times F_{in} \times F_{in} \tag{3}$$

In this research, since we have 20 pooled feature maps $(N = 20)$, each with dimension of 10×10 $(F_{in} = 10)$, then, our flatten layer comprise of 2000 nodes.

Fully Connected Layer with ReLU Function: Fully Connected (FC) layers- as name implies- are those layers where all the inputs from one layer are connected to every activation unit of the next layer [22]. Commonly, FC layers are located as the last few layers of any CNN. Therefore, this module is responsible to compile the high level features extracted by previous layers (convolutional and pooling layers) into a reduced form of low level features in which they can be used by the classifier located at the output layer to provide classification probabilities. In this research, we have developed the FC layer with 200 neurons connected with 2000 nodes of the flattened (FL) layer which provide a layer complexity reduction by $10:1$. As the inputs pass from the units of FL layer through the neurons of FC layer, their values are multiplied by the weights and then pass into the employed activation function (normally ReLU function) just in the same way as in a the classical NN (i.e. shallow NN). Thereafter, they are forwarded to the output classification layer where each neuron expresses a class label. Note that, FC layer also goes through a backpropagation [24] process to determine the most accurate values of its trainable parameters $(weights\ W_{FL} \times W_{FC} = 2000 \times 200)$.

Output Layer with SoftMax Function: This module is responsible to provide/predict the correct classification for each evaluated sample record of the utilized IoT attacks-dataset. Here we are providing two types of classification including the binary-classifier (normal or anomaly) and the multi-classifier (normal, DoS, Probe, R2L, U2R). The data points received from the 200 neurons of the FC layer $(A_1, A_2, \ldots, A_{200})$ are fully connected with the five neurons $(C_1, C_2, C_3, C_4, C_5)$ of the output classes $(j = 5vectors)$ through the transposed weight connections (W_j^T). This can be achieved algebraically as follows:

$$\underline{C} = \underline{W}_j^T.\underline{A} = \begin{bmatrix} \underline{W}_1^T \\ \underline{W}_2^T \\ \underline{W}_3^T \\ \underline{W}_4^T \\ \underline{W}_5^T \end{bmatrix} \begin{bmatrix} A_1 \\ A_2 \\ A_3 \\ \vdots \\ \vdots \\ \vdots \\ A_{198} \\ A_{199} \\ A_{200} \end{bmatrix} = \begin{bmatrix} C_1 \\ C_2 \\ C_3 \\ C_4 \\ C_5 \end{bmatrix} \quad \textit{Where: } \underline{W}_1^T, \underline{W}_2^T, \underline{W}_3^T, \underline{W}_4^T, \underline{W}_5^T \textit{ are vectors of } 1 \times 200 \tag{4}$$

Note that, the output layer also goes through a backpropagation process to determine the most accurate values of its trainable parameters $(weights\ W_{FC} \times W_{out} = 200 \times 5)$. The last layer of the neural network is a *Softmax* layer which has similar number of nodes as the output layer. *Softmax* normalizes the output into a probability distribution on classes [22]. Specifically, *Softmax* assigns numerical probability values for every class at the output layer where these probabilities should sum up to 1.0 (following a probability distribution). Given an input a vector (***x***) of (***K***) real numbers and (***i***) defines the index for the input values, then, SoftMax function $\sigma : \mathbb{R}^k \mapsto \mathbb{R}^k$ is defined as follows:

$$\sigma(\boldsymbol{x})_i = e^{x_i} / \sum\nolimits_{j=1}^{K} e^{x_i} - \boldsymbol{for}\, \boldsymbol{i} = 1, 2, 3, \ldots, \boldsymbol{K}\, \boldsymbol{and}\, \boldsymbol{x} = (\boldsymbol{x}_1, \boldsymbol{x}_1, \ldots, \boldsymbol{x}_K) \in \mathbb{R}^k \quad (5)$$

3.4 System Integration

In this section, we integrate all the aforementioned subsystems and modules by Putting-It-All-Together to come up with complete system architecture of our IoT-IDCS-CNN. Figure 4 illustrates the top view architecture of the integrated system as a feedforward *CovNet* network based IoT attack detection system.

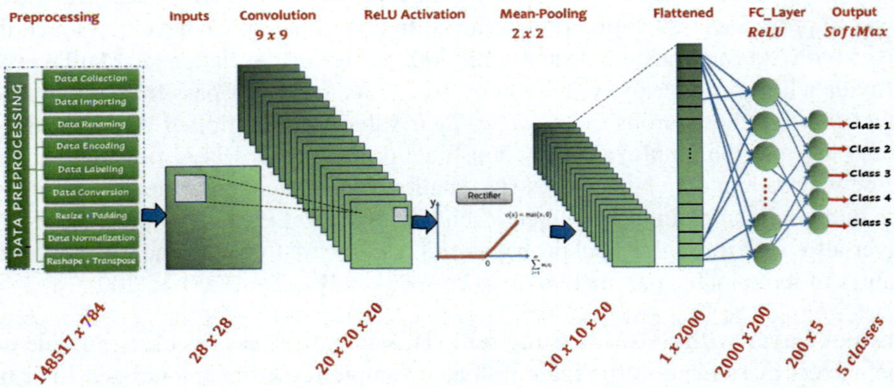

Fig. 4. Top view architecture of the proposed IoT-IDCS-CNN.

According to the system architecture, after data preprocessing stages and using the 28 × 28 input matrix, we constructed 784 (=28 × 28) input nodes. To extract features of the input data, the network encompasses a deep convolutional layer involving a depth of 20 convolution filters of size (9 × 9). Thereafter, the results of the convolutional layer passes via ReLU activation function which followed by the subsampling operation of the pooling layer. The pooling layer utilizes the average pooling method with 2 × 2 submatrices. The pooled features are then flattened to 2000 nodes. The classification/detection neural network comprises the single hidden fully connected (FC) layer and the output classification layer. This FC layer comprises 200 nodes along with ReLU activation function. Since our system requires the classification of the data into 5 classes, therefore, the output layer is implemented with 5 nodes with SoftMax activation function. The next table, Table 2, recaps the final integrated *CovNet* based system for IoT attacks detection.

Moreover, the life cycle for the packet traffic received at the IoT gateway is provided in Fig. 5 below. The input layer takes the encoded features generated from FE subsystem

Table 2. Summary of the developed CovNet for IoT attacks detection/classification system.

Layer	Comment	Trainable parameters
Preprocessing	148517 Sample each (28×28)	–
Input	28×28 nodes (784 nodes)	–
Convolution	20 convolution filters (9×9) + ReLU	W_{Con} $(9 \times 9 \times 20)$
Pooling	Mean pooling (2×2)	–
Flattening	2000 nodes	–
Fully Connected	200 nodes + ReLU	W_{FCL} (2000×200)
Output	5 nodes (or 2 nodes) + SoftMax	W_{Out} (200×5)

in order to be trained at the CNN which update the training parameters and generate the least cost/loss value (error) with optimal accuracy. The output layer employs the SoftMax classifier which is used to classify the data using two classification techniques include: binary classification technique which provides two categories (normal vs anomaly) and the multi-classification technique which provides five categories (normal, DoS attack, Probe attack, R2L attack, U2R attack).

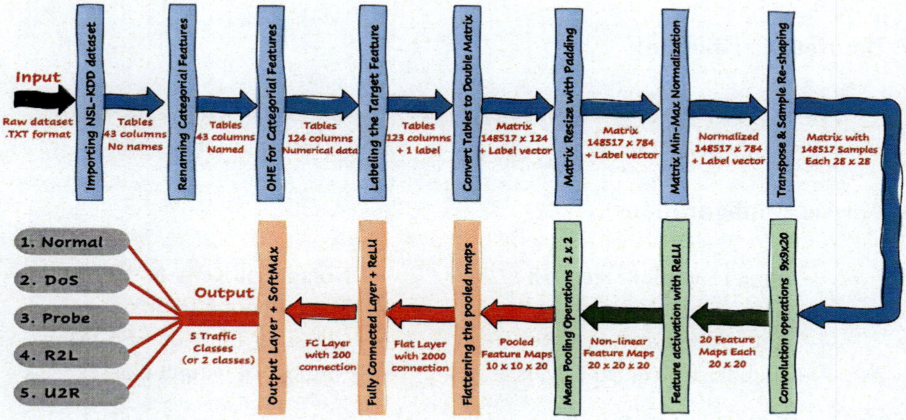

Fig. 5. Comprehensive view of the computation process IoT-IDCS-CNN.

4 Simulation Environment

To implement, verify, and validate the proposed IoT attacks detection and classification system ($IoT - IDCS - CNN$), the training and testing were performed on the NSL-KDD dataset involving the key attacks for IoT communication. The classifier model was determined to have either two classes (binary attack detection) or five classes (multi-attack classification). The proposed system was implemented in MATLAB 2019a. To evaluate the system performance, experiments were performed using a high-performance computing platform utilizing the power of central processing unit (CPU) and graphical processing unit (GPU) with Multicore structure of NVIDIA GeForce® Quadro P2000 Graphic card. The specifications for the workstation used in development, validation & verification are provided via Table 3.

Table 3. The system develompment and validation environment.

System unit	Specifications
Processor Unit (CPU)	Intel Core I9-9900 CPU, 8 Cores, @ 4900 MHz
Graphics Card (GPU)	NVIDIA Quad P2000@1480 MHz, 5 GB Mem, 1024 CUDA Cores
Cache Memory ($)	16 MB Cache @ 3192 MHz
Main Memory (RAM)	32 GB DDR4 @ 2666 MHz
Operating System (OS)	64 bit, Windows 10 Pro
Hard Disk Drive (HD)	SATA 1 TB Drive + 256 GB SSD

Besides, the experimental setup for training/testing model has been configured as follows:

- **Dataset Distribution:**
  - 85% of the dataset used for training (i.e., ~128500 data sample records).
  - 15% of the dataset used for testing (i.e., ~20000 data sample records).

- ***CovNet* Configurations:**
  - Input (Sample) Size = 28 x 28.
  - Conv. Kernel Size = 9 x 9.
  - Activation function = ReLU.
  - Number of Hidden Layers = 5.
  - Number of Kernels = 20.
  - Mean Pooling filter size = 2 x 2.
  - Classifier function= SoftMax.
  - Number of Output classes = 2 or 5.

- **Model Optimization Configurations:**
  - Optimization Algorithm = Mini Batch Gradient Descent (find minimum loss).
  - *Mini_batch_size* = 50, Momentum factor (β) = 0.95, learning rate (α) = 0.05.
  - Momentum updates = $Mom_{Con}[9 \times 9 \times 20]$, $Mom_{FCL}[2000 \times 200]$, $Mom_{Out}[200 \times 5]$.
  - All Momentum updates were initialized using ZEROS matrices ($zeros(size)$).

- **Training Model Configurations:**
  - Training technique = back-propagation with momentum (to update weights).
  - Trainable weights = $W_{Con}[9 \times 9 \times 20]$, $W_{FCL}[2000 \times 200]$, $W_{Out}[200 \times 5]$.

- Backprop. Derivatives = $dW_{Con}[9 \times 9 \times 20]$, $dW_{FCL}[2000 \times 200]$, $dW_{Out}[200 \times 5]$.
- The number of epochs = 100 and the number of iterations per epoch = ~2500.
- All trainable weights were initialized using random number generator (rand).
- All backpropagation derivatives were initialized using ZEROS matrices.

- **Weight update policy:**
  - $dW_{Con} = \frac{dW_{Con}}{Mini\_batch\_size}$, $dW_{FCL} = \frac{dW_{FCL}}{Mini\_batch\_size}$, $dW_{Out} = \frac{dW_{Out}}{Mini\_batch\_size}$
  - $\text{Mom}_{Con} = \alpha * dW_{Con} + \beta * \text{Mom}_{Con}; \rightarrow W_{Con} = W_{Con} + \text{Mom}_{Con}$
  - $\text{Mom}_{FCL} = \alpha * dW_{FCL} + \beta * \text{Mom}_{FCL}; \rightarrow W_{FCL} = W_{FCL} + \text{Mom}_{FCL}$
  - $\text{Mom}_{Out} = \alpha * dW_{Out} + \beta * \text{Mom}_{Out}; \rightarrow W_{Out} = W_{Out} + \text{Mom}_{Out}$.

5 Results and Discussion

To verify the effectiveness of the proposed system in compliance with its intended functionalities and missions, we have evaluated the system performance using the recommended testing dataset in terms of the classification accuracy, classification error percent and the classification time as follows:

$$\textit{ClassificationAccuracy}(\%) = \frac{\textit{CorrectlyPredictedSamples}}{\textit{NumberofTestingSamples}} \times 100\% \quad (6)$$

$$\textit{Classification Error}(\%) = \frac{\textit{Incorrectly Predicted Samples}}{\textit{Number of Testing Samples}} \times 100\% \quad (7)$$

$$\textit{Classification Time}(ms) = \sum_{i=1}^{No.Runs} \textit{Execution time } (i) \times \frac{1000}{No.Runs} \quad (8)$$

The plot for the overall testing classification accuracy and overall classification loss (classification error) comparing the performance of the binary-classifier (2-Classes) and the multi-classifier (5-Classes) obtained during the validation process of NSL-KDD dataset are illustrated in Fig. 6. According to the figure, at the beginning and after one complete pass (epoch) of testing process, both classifiers showed relatively low classification accuracy proportions with 85% and 79% registered for 2-Class classifier and 5-Class classifier, respectively. Thereafter, both classification accuracy curves begin to roughly be increasing in a stable tendency while testing epochs proceeds with faster and higher ceiling level obtained for the classification accuracy of 2-Class classifier. After training the system for 100 epochs, the system was able to record an overall testing accuracy proportions of 99.3% and 98.2% for 2-Classs classifier and 5-Classs classifier, respectively, for the given testing dataset samples. Conversely, it can be clearly seen that both classifiers showed relatively high classification error proportions at the beginning of the testing process with 15% and 21% registered for 2-Class classifier and 5-Class classifier after one testing epoch, respectively. Thereafter, both classification error rates started to systematically decline while the binary classifier progresses with faster threshold achieving 0.7% of incorrect prediction proportion (classification error percentage). However, the classification error rate proportion for the multi-classifier has saturated with less than 2.0% of incorrect prediction. This range of classification error of both classifiers (0.7%–1.8%) is permitted to avoid underfitting or overfitting from the training loss (~0.0%) and training accuracy (~100%) and thus provided high-accuracy classification performance.

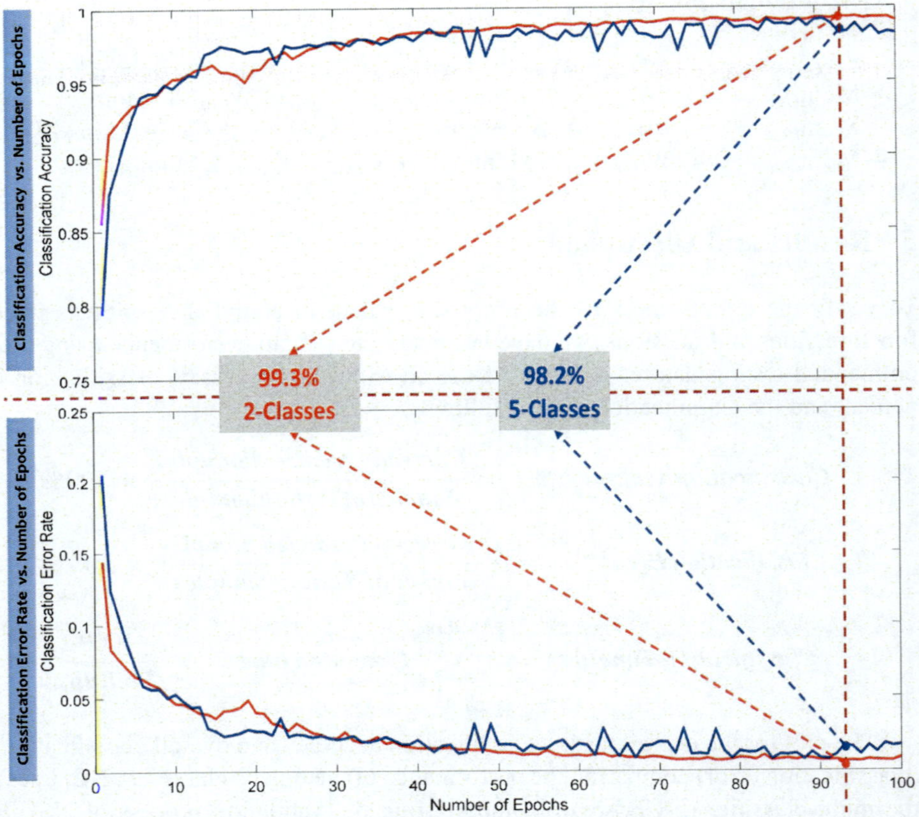

Fig. 6. Testing Detection/Classification Accuracy/Error Rate vs. number of Epochs.

Moreover, we have analyzed the time required to perform attack detection or classification for one IoT traffic sample. To obtain accurate and precise results, we have run the validation test for 500 times and then computed the time statistics for detection and classification. Figure 7 shows the detection/classification time performance for the proposed model (either 2-Class or 5-class classifier). According to the figure, the time required to detect/classify one sample record ranges from (*Min* $\approx$ 0.5662 ms) to (*Max* $\approx$ 2.099 ms) with average time of (*Mean* $\approx$ 0.9439 ms) recorded for the 500 simulation runs. This average time (*around* 1 ms) is very useful for the system to run in dynamical environment such as the real time IDS applications. Finally, to gain more insight on the advantage of the proposed method, we benchmark IoT-IDS-CNN classification system by comparing its performance with other state-of-art machine learning based intrusion/attacks detection systems in terms in terms of classification accuracy metric. For better and more reasonable evaluation, we have selected the related researches that employs machine learning techniques for intrusion/attacks detection/classification for the NSL-KDD dataset (the same used by our system) to be compared with our proposed IoT-IDS-CNN. Therefore, we summarize the classification accuracy metric values for related state-of-art research's in the following table, Table 4, in chronological order. Accordingly, it can be obviously noticed, that the proposed IoT-IDS-CNN model has improved the cyber-attacks classification accuracy of other ML-IDS models by an improvement factor (IF) of (~1.03 – 1.25).

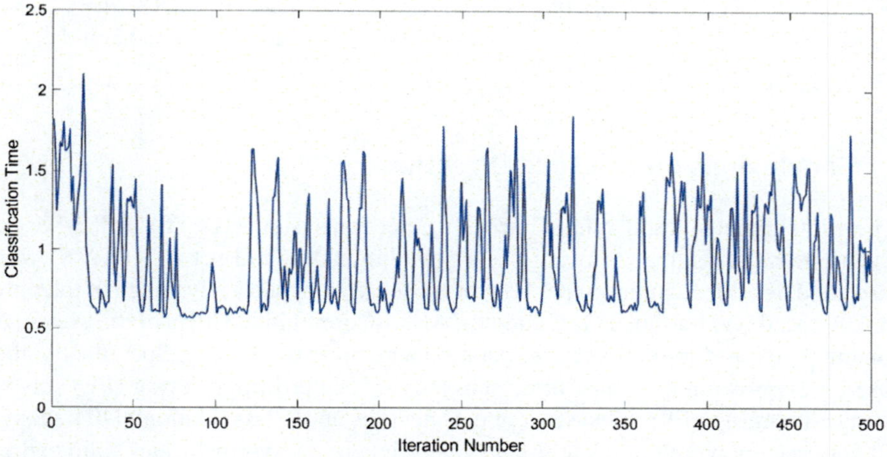

Fig. 7. Run time performance of IoT Traffic classification over 500 simulation runs.

Table 4. Comparison with state-of-art ML-IDS employing same dataset.

Research	Method	Accuracy	Remark	IF %
K. Taher et al. 2019 [6]	ANN with SVM Classifier	≈83.7%	3-classes, with 2 hidden layers and used only 35-features	1.173
X. Gao et al. 2019 [7]	Deep NN with ensemble voting	≈85.2%	5-classes, 3-methods: Decision Tree, Random Forest, K-Nearest	1.152
S. Sapre, et al. 2019 [8]	Different ML-IDS techniques	≈78.5%	5-classes, with 2 hidden layers and Naïve Bayes Classifier	1.251
Chowdhry, 2017 [9]	DNN with SVM + K-Nearest	≈94.6%	5-classes, 6 hidden layers, used only 35-features, 1-NN Classifier	1.038

(*continued*)

Table 4. (*continued*)

Research	Method	Accuracy	Remark	IF %
Javaid et al. 2016 [10]	Deep Autoencoders (DAE)	≈88.4%	2-classes, with 5 hidden layers and SoftMax classifier	1.123
Yadigar, et al. 2016 [11]	DNN with Extreme Learning Machine(ELM)	≈91.7%	2-classes, Hybrid feature selection: (consistency subset evaluation + DoS characteristic features), used only DoS attacks, 17 Features	1.080
Proposed Method	CNN: IoT-IDS_*CNN*	≈98.2% ≈99.3%	2-classes 5-classes CNN, 5 hidden layers and SoftMax classifier	-

6 Conclusions and Future Directions

A new intelligent and self-reliant intrusion detection system (IDS) for IoT cyber-attacks detection and classification using deep convolutional neural network (CNN) was proposed and developed in this paper. The proposed system takes advantage of the powerful CUDA based Nvidia-Quad GPUs and the Multi-Core I9 Intel CPUs of the system development platform for improved computation and faster processing. Specifically, the proposed system is composed of three subsystems: Feature Engineering (FE) subsystem, Feature Learning (FL) subsystem and Detection and Classification (DC) subsystem. All subsystems were completely developed, integrated, verified, and validated in this research. Because of the use of CNN based design, the proposed system was able to detect the slightly mutated attacks of IoT networking (represented collectively by NSL-KDD dataset which includes all the key attacks in the IoT computing) with detection accuracy of **99.3%** of normal or anomaly traffic, and classify the IoT traffic into five categories with classification accuracy of **98.2%**. Such results of the implemented system outperform several similar up-to-date state-of-art research results in the same area of study. In future, we will consider the inclusion of more detailed description of our system modules and the machine learning algorithms. Also, we will perform and incorporate more comprehensive performance analysis to gain more insight about the system efficiency such as the confusion matrix to analyze the attacks' detection True/False Positives and the True/False Negatives and other evaluation metrics including the Precision, Recall, F-Score Metric, False Alarm Rate. Moreover, we will provide more robust validation for our system using K-Fold Cross Validation as well as the comparison of our system with other State-of-Art ML-IDS employing different dataset.

References

1. Alrawais, A., Alhothaily, A., Hu, C., Cheng, X.: Fog Computing for the Internet of Things: Security and Privacy Issues. IEEE Internet Comput. **21**(2), 34–42 (2017)
2. Chiti, F., et al.: Context-aware wireless mobile automatic computing and communications: research trends & emerging applications. IEEE Wirel. Commun. **23**(2), 86–92 (2016)

3. Silva, B.N., Khan, M., Han, K.: Internet of things: a comprehensive review of enabling technologies, architecture, and challenges. IETE Tech. Rev. **35**(2), 205–220 (2017). https://doi.org/10.1080/02564602.2016.1276416
4. Mahmoud, R. et. al. IoT security: current status. In: 10th International Conference for Internet Technology and Secured Transactions (ICITST), pp. 336 341 (2015)
5. Paar, C., Pelzl, J.: Understanding Cryptography. Springer, Berlin https://doi.org/10.1007/978-3-642-04101-3 (2010)
6. Caspi, G.: Introducing Deep Learning: Boosting Cybersecurity with An Artificial Brain. Informa Tech, Dark Reading, Analytics http://www.darkreading.com/analytics. (2016)
7. Taher, K.A., Jisan, B.M., and Rahman, M.M.: Network intrusion detection using supervised machine learning technique with feature selection. In: IEEE International Conference on Robotics, Electrical and Signal Processing Techniques, pp. 643–646 (2019)
8. Gao, X., Shan, C., Hu, C., Niu, Z., Liu, Z.: An adaptive ensemble machine learning model for intrusion detection. IEEE Access **7**, 512–521 (2019)
9. Sapre, S., Ahmadi, P., Islam, K.: A Robust Comparison of the KDDCup99 and NSL-KDD IoT Network Intrusion Detection Datasets Through Various Machine Learning Algorithms. arXiv:1912.13204v1 [cs.LG]. (2019)
10. Chowdhury, M. et. al.: A few-shot deep learning approach for improved intrusion detection. In: IEEE 8th Annual Ubiquitous Computing, Electronics and Mobile Communication Conference (UEMCON), New York, pp. 456–462, https://doi.org/10.1109/uemcon.2017.8249084. (2017)
11. Niyaz, Q., Sun, W., Javaid, A.Y., Alam, M.: Deep learning approach for network intrusion detection system. In: ACM 9th EAI International Conference on Bio-inspired Information and Communications Technologies, New York (2016)
12. Yusof, A.R. et.al.: Adaptive feature selection for denial of services (DoS) attack. In: IEEE Conference on Application, Information and Network Security (AINS), Miri 2017, pp. 81–84 https://doi.org/10.1109/ains.2017.8270429. (2017)
13. Canadian Institute for Cybersecurity (CIS). NSL-KDD Dataset. https://www.unb.ca/cic/datasets/nsl.html. Retrieved on. (2019)
14. A. Ozgur and H. Erdem. A review of KDD99 dataset usage in intrusion detection and machine learning between 2010 and 2015. PeerJ Preprints 4:e1954v1. (2016)
15. C. Kolias, G. Kambourakis, A. Stavrou and J. Voas. DDoS in the IoT: Mirai and Other Botnets," Journal of Computers, vol. 50 (7), pp. 80–84, https://doi.org/10.1109/mc.2017.201. (2017)
16. C. Ambedkar, V. Kishore Babu. Detection of Probe Attacks Using Machine Learning Techniques. International Journal of Research Studies in Computer Science and Engineering (IJRSCSE), Vol2, No. 3, Pp 25–29. (2015)
17. P. Pongle and G. Chavan. A survey: Attacks on RPL and 6LoWPAN in IoT," 2015 International Conference on Pervasive Computing (ICPC), Pune, 2015, pp. 1–6. https://doi.org/10.1109/pervasive.2015.7087034. (2015)
18. Bengio, Y., Courville, A., Vincent, P.: Representation learning: a review and new perspectives. IEEE Trans. Pattern Anal. Mach. Intell. **35**(8), 1798–1828. arXiv:1206.5538. https://doi.org/10.1109/tpami.2013.50. PMID 23787338. (2013)
19. Sarkar, D.J.: Understanding Feature Engineering. Towards Data Science. Medium. https://towardsdatascience.com/tagged/tds-feature-engineering. (2018)
20. Brownlee, J.: A Gentle Introduction to Padding and Stride for Convolutional Neural Networks. Deep Learning for Computer Vision, Machine Learning Mastery. https://machinelearningmastery.com/padding-and-stride-for-convolutional-neural-networks (2019)
21. Kalay, A.: Preprocessing for Neural Networks - Normalization Techniques. Machine Learning, Github.IO. https://alfurka.github.io/2018-11-10-preprocessing-for-nn. (2018)
22. Li, F.: CS231n: Convolutional Neural Networks for Visual Recognition. Computer Science, Stanford University, http://cs231n.stanford.edu. (2019)

23. Brownlee, J.: A Gentle Introduction to the Rectified Linear Unit (ReLU). Deep Learning for Computer Vision, Machine Learning Master. https://machinelearningmastery.com. (2019)
24. Kim, P.: MATLAB Deep Learning with Machine Learning. Apress, Neural Networks and Artificial Intelligence (2017)

On Federated Cyber Range Network Interconnection

Adamantini Peratikou[1](✉), Constantinos Louca[1], Stavros Shiaeles[2], and Stavros Stavrou[1]

[1] Faculty of Pure and Applied Sciences, Open University of Cyprus, 2220 Nicosia, Cyprus
adamantini.peratikou@ouc.ac.cy

[2] School of Computing, Faculty of Technology, University of Portsmouth, Buckingham Building, Lion Terrace, Portsmouth PO1 3HF, UK

Abstract. Cyber Ranges exist to enable hands on training within realistic ICT infrastructures in a sandboxed environment, to investigate attack and defense strategies and to assess the resilience of the infrastructures. To fully exploit their capabilities one has to have access to multi domain exercises, which may combine ICT, naval, electrical grid, telecom or other relevant infrastructures. It can become obvious that no single organization can easily own or sustain a multi domain cyber range and that there is a need to connect multi domain Cyber Ranges from different organizations together. This paper focuses into analyzing the current state of the art on the federation of Cyber Ranges, by focusing on the federated network interconnection. Various methods for interconnecting distributed Cyber Ranges into a single federated Cyber Range are being discussed and their network performance impact is evaluated. VPNs are widely used to interconnect networks together due to their relative low cost and simplistic nature, however, performance of the network must be accounted, alongside the flexibility the VPNs can provide to support multiple scenarios in a multi domain distributed federated Cyber Range. This work focuses on the performance comparison of IPsec and Virtual Tunnels.

Keywords: Cyber Ranges · Federation · Interconnection · VPN · IPsec · OpenVPN · Virtual tunnels · Cyber security

1 Introduction

In an increasingly networked connected world, where successful cyber-attacks can have disastrous effects, which are not retained to the targeted system but can also affect multiple sectors of the economy, it is of crucial importance that cyber security practitioners are trained in realistic multi domain training environments. This would allow them to hone their skills and enable them to have cyber-attack response experience without actually being exposed to a real attack [1]. Given that it is inadvisable to use a live system for training purposes, or to take it offline so that users can train on it, the need arises for creating Cyber Ranges (CRs). CRs of this kind tend to be focused on that area of expertise

B. Ghita and S. Shiaeles (Eds.): INC 2020, LNNS 180, pp. 117–128, 2021.
https://doi.org/10.1007/978-3-030-64758-2_9

specific to the entities running them, and while they can be quite comprehensive in the study of their own subsection of cyberspace, they are often limited to that subsection [2].

For example, while a CR focusing on testing the vulnerability of an e-banking system can be very helpful in allowing researchers to draw conclusions on securing e-banking systems, it can offer no insights as to the possible consequences of a particular threat to another system, for example Electrical Grid oversight, or Air Traffic Control [3]. This is to a large extent due to the fact that the threats to one type of system are erroneously perceived as unrelated or irrelevant to another, as well as to the fact that testing for the effects of these threats on a system requires very specific information on the functioning of that particular system. Different systems have different services, resources and requirements, and in order to draw accurate conclusions, one needs to be able to study a threat across a wider range of systems, including some that one would not normally have access to, or expertise on.

The Federation of CRs attempts to fill this gap, by pooling together different CRs form similar or different domains, and testing for the effects of a particular threat across a more widespread selection of targeted systems, emulating therefore the true nature of the interconnected world. This paper examines both the theoretical soundness of pooling resources in this manner, as well the technical methodology required to do so.

The first section of this paper provides some possible techniques of implementing the interconnection of federated CRs by overviewing different technologies. The second section presents relevant performance metrics. The third section analyzes the requirements and proposes an interconnection architecture to satisfy those requirements. The final section provides some preliminary results obtained via performance tests to identify the appropriateness of the solution.

2 Related Works

The European Defense Agency currently works on a project for creating a federated CR environment [2] where EU Member States will federate their national CRs and improve their respective Cyber Defense training capabilities. The congressional record proceedings of the 113th congress session of the United States [4], documents an attempt to interconnect individual CRs together in a distributed manner. Another example is the EU-funded project ECHO and the CyberSec4Europe that upon completion will analyze requirements, specifications, and use cases for federation of CRs [5]. Also, the EU-funded project Foresight aims to provide a multi-domain geographically distributed Federated Cyber Range [6]. A number of CRs are currently in use or being developed and provided by third party organizations across governments, academia and private sector. CRs provided by universities are predominantly used for education and training, most of them have just emerged in the last few years. Currently the most established CR in a university setting is in Michigan and is being governed by 12 public universities, named the Merit Network [7]. The CR has been available since 2012, it has four physical locations and incorporates a virtual training environment called "Alphaville" to evaluate cybersecurity skills. Another example is the CR called "The DETERLab" at the University of Southern California [7]. CRs provided by industry are primarily used for the training of cyber security professionals or used from other companies for finding vulnerabilities in their own infrastructures [8].

While several governments are investing in CRs, and research is currently conducted to interconnect CRs together, the information that is publicly available is limited. Substantial research interest has been dedicated to technologies that enable the interconnection of CRs, but there is a significant gap in the literature when it comes on interconnecting CRs with the use of VPNs.

Various VPN technologies exist that are based on Point-to-Point Tunneling Protocol (PPTP), IP Security standard protocol (IPsec) or SSL (Secure Sockets Layer) technology [9]. PPTP are not generally preferred due to security issues, arising from the simplicity of the protocol [10]. For interconnecting CRs in a federated environment, IPsec and SSL/TLS based VPN tunnels will be compared. IPsec and SSL are a set of cryptographic protocols that provide secure communication. IPsec protocols include AH, ESP, IKE, ISAKMP/Oakley, and transforms [11]. IPsec connection starts with a two phase handshake and when it is completed, arbitrary traffic can be sent via an encrypted tunnel [12].

By decoding the current trend and the need for realistic training and evaluation, we can foresee more and more CRs being developed in the forthcoming years, where the individual CR capabilities can be extended by interconnecting individual CRs together in a federated environment.

3 CR Interconnection Framework

3.1 Framework Metrics

A federated CR interconnection environment above all must be reliable and efficient. To assess the effectiveness of such environment, or framework, a set of measurements need to be satisfied. Various metrics span across research communities and agencies tasked with cybersecurity [13]. A general framework without any detailed recommendations, ITIL, is used by organizations for service quality management, while COBIT is used by several organizations as a measurement technique in security areas. Considering that the federation of CRs is still in an evolving state, we are proposing a set of minimum measures that need to be satisfied.

Scalability: Due to the emerging nature of CRs the interconnection framework should be scalable enough to support an increasing number of CRs to be connected in the federated environment.

Fault Tolerance: The interconnection of the federated CRs has to be resilient and able to intelligently handle faults as the VMs under the CRs toggle between offline and online states and participate in exercises across geographically distributed multi domain CRs.

Availability: Availability is defined as the probability of receiving a fulfilled service request in an acceptable time [14] The interconnection of CRs has to be up at all times and provide the service requested by CR users within a reasonable time.

Performance: For the performance metrics, a set of attributes must be taken into consideration such as Response time, throughput, CPU load, memory and network usage. Also depending on the interconnection technology, whether VPN or IPsec, the number of simultaneous users that can be supported is also required.

Security: Security can be interpreted in terms of confidentiality, integrity and availability which are considered difficult to measure but are assumed that are handled by a well-accepted secure protocol like IPsec etc.

3.2 CR Federation Requirements

A realistic federated Cyber Range training environment should allow the interconnection of any CR domain environment through its network architecture. This capability enables an increased number of end-users, belonging to different domains, to be trained in a multi-domain environment. By implementing federated gateways, and relevant functionality, a realistic federated Cyber Training environment between remotely distributed platforms can be established. Federated gateways can be realised through the use of Virtual Private Networks (VPNs). The addressing schemes in such methods must be agreed upon the CRs so that each CR has the appropriate address space in the subnets that will be used in the distributed federated scenarios.

It is envisioned that organisations that own a CR or cyber training environment, will use what we call a federated gateway, to connect their own platform or resources to other CRs running their own federated gateway. This interconnection will allow the combination and sharing of remote resources, to form a very large scale distributed federated CR.

3.3 Interconnection Requirements

The interconnection architecture considers the various cyber-ranges and connects them using a centralized federated gateway to form a unified platform. The architecture should be designed in such a way to allow the provision of management, deployment, and usage of VMs and Cyber scenarios of interest. With that in mind the remote gateways should have a centralized master federation point of reference to serve as the management for the whole federation platform environment e.g. to monitor and manage the status of the interconnection links at any given time and take corrective action in case of a CR link failure.

The requirements of the federated gateways are as follows:

- To support ethernet TCP/IP communication
- Federated gateways will be based on VPN technologies
- Point to multipoint, or mesh connection
- Monitoring service/process to constantly monitor connectivity between CRs

3.4 Interconnection Techniques

A federated interconnection can be implemented at layer 1, layer 2 or layer 3. Interconnecting at layer 1 is very challenging for supporting complex scenarios and exercises and has practical cost and implementation implications. On the other hand, layer 2 and layer 3 interconnectivity are less costly to establish and have the potential to grow to an ad-hoc federation of geographically distributed interconnected CRs across the world. Virtual Private Networks (VPNs) allow secure connections to a private network, across any public network, and replace the need to physically lay private cables over long distances [15]. The most appropriate method to allow federation between CRs is with the use of VPNs due to their proven cost effectiveness.

For the accomplishment of layer 2 VPN services, Fig. 1, the traffic on layer 2 VPN is transported to the service provider network by MPLS, which on the sending and receiving ends it gets transformed back to Layer 2 format. Different Layer 2 formats can be configured at the sending and receiving ends. The security and privacy of an MPLS Layer 2 VPN are equal to those of an ATM or Frame Relay VPN [16]. The service provisioned with Layer 2 VPNs is also known as Virtual Private Wire Service (VPWS). Generally, on a layer 2 VPN the routing is done on the customers routers, therefore CR providers routers. Those routers must choose the correct circuit to send the traffic to. In the provider edge, the routers receiving the traffic send it across the service provider network. For a Layer 2 VPN, the clients routers need to be configured to carry all layer 3 traffic, while the routers on the service provider only require to know the amount of traffic the VPN layer 2 will carry, to carry that traffic between the client using layer 2 VPN interfaces [17]. The policies must be configured on the provider routers. The clients are only required to know which VPN interfaces connect to which of their own sites. Figure 1 illustrates a Layer 2 VPN in which each site has a VPN interface linked to each of the other client sites.

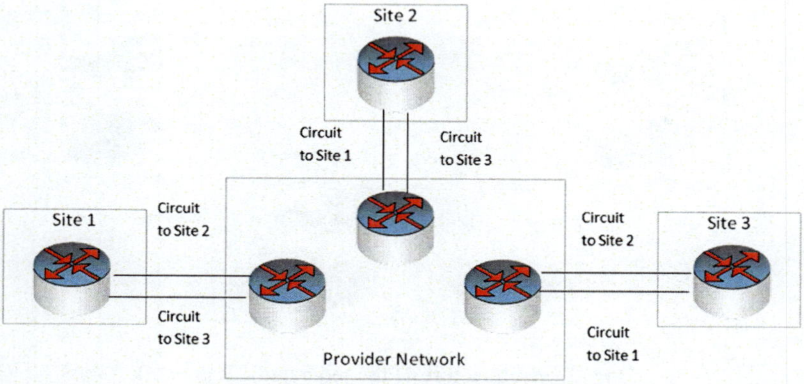

Fig. 1. Layer 2 VPN Connecting Routers

While layer 2 VPN can connect CRs in one big federation environment, there would be a need for additional control mechanisms for monitoring and managing those CRs. While planning this kind of connection, an input from every range architecture is needed.

Different IP addressing and VLAN schemes, environment isolations and implementation of monitoring and situational awareness tools must be agreed among the interconnected CRs. To avoid this complexity and provide a more controlled CR environment, where each CR has more security control over their own interconnected CR, it was decided that a layer 3 VPN connection will be implemented and tested for the federated gateways. The CRs need to exchange resources with the use of site to site VPNs. The options for layer 3 site to site VPNs are IPsec and SSL/TLS from OpenVPN. IPsec establishes VPN tunnels through the use of protocol and cryptographic security measures. It supports network-level peer authentication, access control, connectionless integrity, data origin authentication, detection and rejection of replays, confidentiality, and limited traffic flow confidentiality [18]. IPsec uses two modes of operation, the tunnel mode and the transport mode. Tunnel mode is used in site to site communication and it encrypts the header and the payload. The protocols utilized by IPsec to allow traffic security are the Encapsulating Security Payload (ESP) and the Authentication Header (AH). ESP offers integrity data origin authentication and confidentiality. AH is used for packet integrity validation and sender authentication. IPsec connects peers through the Internet Key Exchange (IKE) protocol [19].

OpenVPN utilizes the SSL/TLS protocol for cryptographic elements required by VPN and tunnel creation [20]. OpenVPN instead of accessing the network interface like IPsec, it generates instead a virtual network interface. It encrypts traffic using the same principles for the handshake as IPsec-IKE, and SSL libraries are then used to secure the symmetric tunnel.

3.5 Network Architecture

The high level network architecture of a federated CR is depicted in Fig. 2.

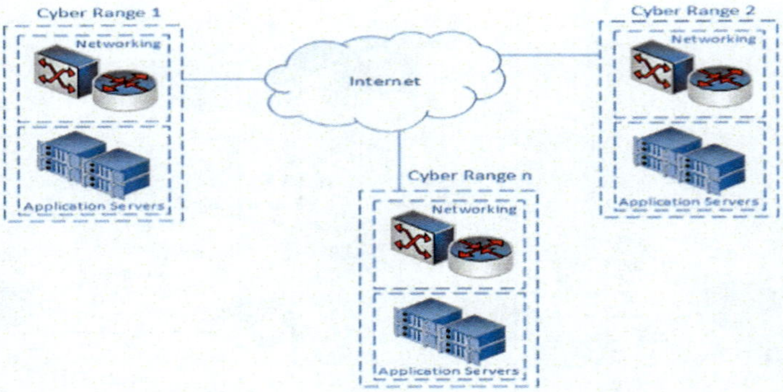

Fig. 2. Architecture of the federated CR network

This federated network serves as a virtual environment where all multi domain cyber security scenarios can be executed. The number of CRs can increase dynamically, in case a new CR vendor is interested in joining the federation.

The proposed interconnection architecture utilizes Layer 3 VPN technologies. VPN technologies that can be used in a federated CR environment consist of the standard remote VPN that will be responsible for routing all the traffic using specific routing policies and site to site virtual tunnels. A site to site setup will allow different networks to be connected using the VPN tunnel. To promote collaboration and accessibility, open source packages are used on federated gateways. An OpenVPN client/server will be used to setup an encrypted remote VPN.

In the point to multipoint configuration, the federated gateway controllers require a host to act as the central federated controller (Fig. 3). Each additional CR location will be set up with a connection profile connected to the central federated controller. Once the server is configured accordingly and the connection profile is installed on the CR's hosting machine, the CR will be connected to the federated network. Then, CRs will be automatically connected to all the resources that the central federated controller provides. This means that the server security controls can be enforced site-to-site, and the resources of the network are accessible at all locations. IPsec site to site requires a specific set of ports, 400 and 4500, to be available by all participating CRs while OpenVPN site to site tunnels requires one port per tunnel [21].

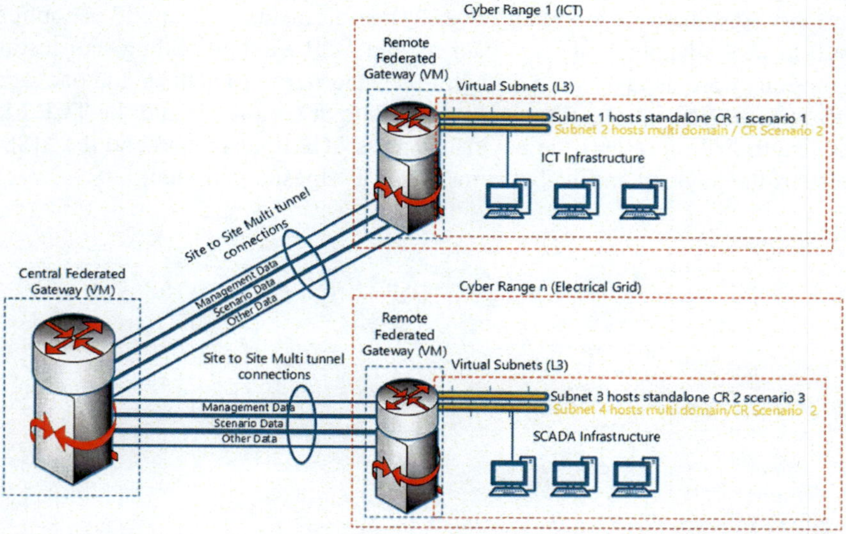

Fig. 3. Proposed Interconnection

Monitoring

A process is required to monitor the federated gateways' controller status of the various CRs at all times. To achieve this a web based application was designed. The application displays the status of each CR that serves as a gateway controller (server), including all current connections (clients). The application works by parsing VPN data from the controller logs. The web based application shows all the relevant information for each VPN server and client. For the server it highlights the VPN mode, status whether connected or disconnected, number of clients, traffic, UP time and local IP of the DHCP server on the federated gateway. For the connected client it illustrates the hostname, the VPN IP & Remote IP, the location of the client, the traffic, the last contacted (last ping) and the uptime. A map is also generated to show the exact location of the deployed CR. The web application can also serve as a simple heartbeat monitor for all gateway clients and server instances.

4 Results

To characterize the network performance of the different CR interconnecting technologies, iperf3 was used. A basic test setup is illustrated in Fig. 4. The iperf3 tool was run through a script that was automatically executed on a predefined time interval. The script collects the IPs of the connected federated gateway controllers (CRs) from the monitoring tool, specified above, and runs iperf3 TCP and UDP tests on each. Because the most common use of iperf3 is to measure the speed of data transfer, the reported results will be the bytes per second of the data transfer rather than the Layer-3 network performance transfer characteristics. To compensate for this, a set of multipliers are being used to include the calculated overhead for specific packet lengths and to convert from iperf3 results to Layer-3 and Layer-2 measurements [22]. Iperf3 is configured by default to take advantage of large packets and TCP window scaling. For testing, specific options were used to force different package sizes. To achieve the packages sizes iperf3 requires configuring flags such as the length option (-l) and the TCP MSS (-M).The length flag represents the payload (minus TCP timestamps) and the MSS option represents the payload and 12-bytes for the TCP timestamp option.

Fig. 4. Performance tool architecture

4.1 Tests Parameters

Throughput tests were run on both VPN technologies and over an unencrypted connection under various packet sizes.

Tests were run over a 1 Gbit/sec ethernet connection. Testing environment includes, two virtualized Debian based routers, one acting as a central federated controller that resides on one CR, and a second instance residing on a different CR, and two virtual machines (VMs). A Debian VM on the first CR site connected to a subnet of the first router (RT1) and an Ubuntu VM the second site connected to a subnet of the second router (RT2). The Debian based routers are running VyOS (a Linux-based network router and firewall distribution) [23]. The network layout and environment infrastructure of IPsec and OpenVPN virtual tunnels were identical, as presented in Fig. 3.

Figure 5 illustrates the TCP throughput obtained when a packet size of 128-1024 is set for both VPN configurations. The VM residing on RT1 subnet acted as a server and the second VM on RT2 subnet as a client as presented in Fig. 4. The throughput is normalized, expressed as the ratio of the throughput obtained from the VPN technology under test, to the throughput of the same unencrypted connection. As expected, the throughput increases proportionally with the packet load. The results show that the throughput of IPsec surpasses by up to 8% the throughput of OpenVPN virtual tunnel for all packet sizes. Figure 6 shows the latency for IPsec, OpenVPN and the unencrypted ethernet connection. IPsec has slightly lower latency than OpenVPN virtual tunnels. Latency for both VPN protocols, as expected, is noticeably higher than the unencrypted ethernet configuration. Results presented in Fig. 5 and Fig. 6 suggests that the performance of the two protocols is comparable.

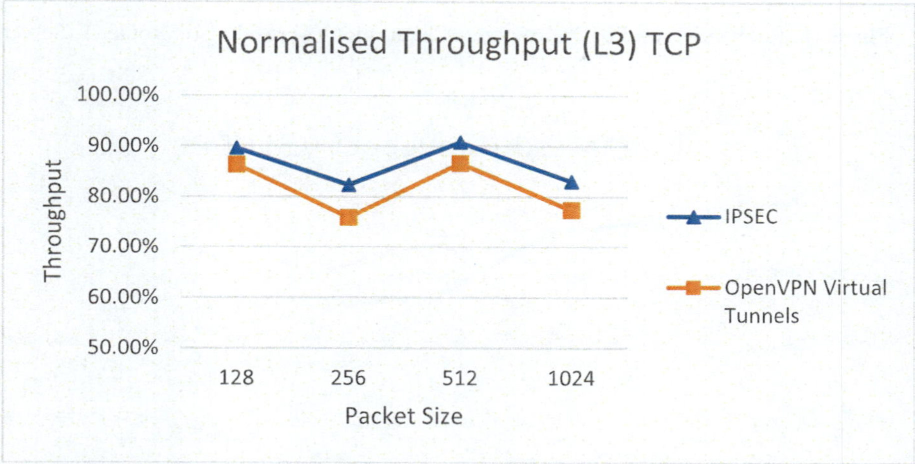

Fig. 5. Normalized Layer 3 TCP Throughput (a VM residing on Router 2 (Client) to a VM residing on Router 1 (Server))

For the same scenario as above, UDP throughput tests were also conducted, shown in Fig. 7. The performance of the two VPN technologies was found to be comparable for packet sizes 128–256, while for packet sizes 512–1024 IPsec performance marginally decreases. The difference between the two VPN configurations is too minimal to be considered as an advantage for one versus the other.

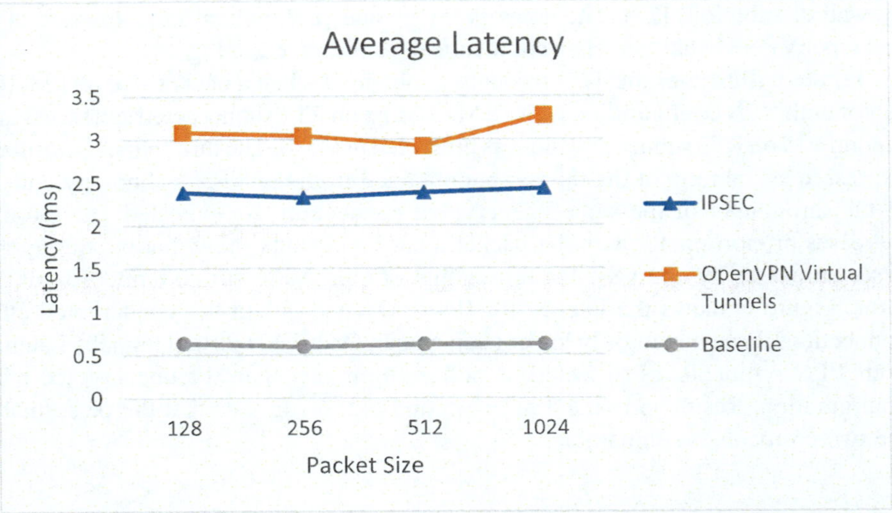

Fig. 6. Latency (VM residing on Router 2 (Client) to a VM residing on Router 1 (Server))

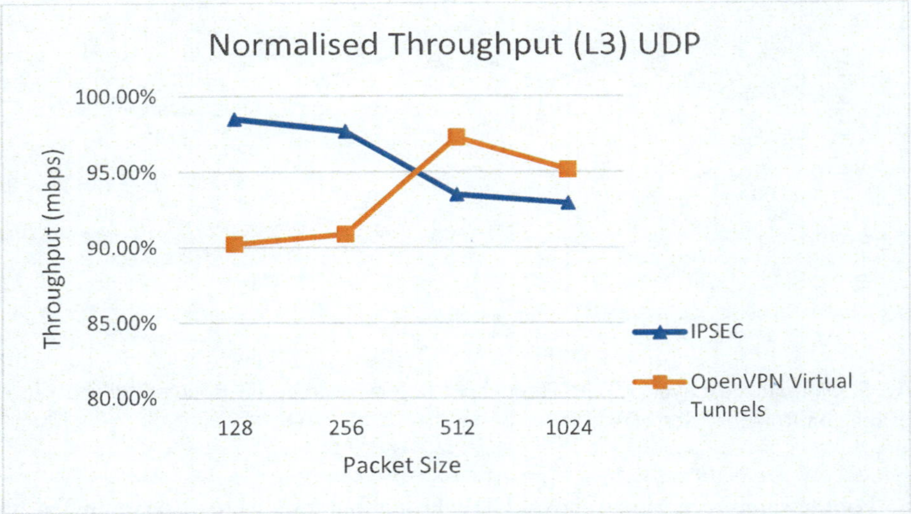

Fig. 7. Layer 3 UDP Throughput (a VM residing on Router 2 (Client) to a VM residing on Router 1 (Server))

5 Conclusions

This paper overviewed and compared possible techniques for realizing a generalised interconnection architecture for federated CRs. The framework performance metrics that need to be taken into account when interconnecting CRs have been presented. Preliminary results obtained through network tests to identify the performance of the protocols involved have been also provided.

Our evaluation encompassed a wide range of packet sizes and identified that IPsec and OpenVPN virtual tunnels behave similarly in all packet sizes with no major differences when it comes to performance. In general, the performance depends on the interconnecting link bandwidth, the speed of the encrypting and decrypting device and the type of cipher in use. The proposed federated interconnection architecture illustrated in Fig. 3 can be implemented using either OpenVPN or IPsec depending on the availability of network ports, with respect to how many CRs are involved in the federation and how many geographically distributed subnets of these CRs are involved in the scenarios under investigation.

Acknowledgement. The authors would like to acknowledge the FORESIGHT project funded by the European Union's Horizon 2020 research and innovation programme (grant agreement: 833673), and the partners on the project.

References

1. Debatty, T., Mees, W.: Building a CR for training CyberDefense situation awareness. In: 2019 International Conference on Military Communications and Information Systems, ICMCIS, pp. 1–6. Budva, Montenegro (2019). https://doi.org/10.1109/icmcis.2019.8842802
2. Ferguson, B., Tall, A., Olsen, D.: National cyber range overview. In: 2014 IEEE Military Communications Conference, pp. 123–128. IEEE (2014). http://doi.org/10.1109/MILCOM.2014.27
3. Ellis, R., Mohan, V.: Rewired Cybersecurity Governance. Wiley, Somerset (2019)
4. Records, C.: Proceedings and debates of the 112th congress. United Stated Government Publishing Office, Washington DC (2012)
5. Directorate General for Internal Policies: Cybersecurity in the European Union and Beyond: Exploring the Threats and Policy Responses, European Parliament (2015)
6. FORESIGHT Project Homepage. https://foresight-h2020.eu. Accessed 10 July 2020
7. Urias, V., Stout, W.M.S., Van Leeuwen, B., Lin, H.: CR infrastructure limitations and needs of tomorrow: a position paper. US (2018). https://doi.org/10.1109/CCST.2018.8585460
8. IBM, IBM Invests $200 M. https://www.enterprisetimes.co.uk/2016/11/17/ibm-spends-200m-cyber-range/. Accessed 12 July 2020
9. Berger, T.: Analysis of current VPN technologies. In: 1st International Conference on Availability, Reliability and Security (ARES 2006), IEEE, pp. 108–115. Vienna (2006). https://doi.org/10.1109/ares.2006.30
10. Kotuliak, P.R, Trúchly P.: Performance comparison of IPsec and TLS based VPN technologies. In: 9th International Conference on Emerging eLearning Technologies and Applications (ICETA), pp. 217–221. Stara Lesna (2011). https://doi.org/10.1109/iceta.2011.6112567
11. Dhall, H., Dhall, D., Batra, S., Rani, P.: Implementation of IPsec protocol In: 2nd International Conference on Advanced Computing & Communication Technologies 2011, pp. 176–182. Rohtak, Haryana (2012). https://doi.org/10.1109/acct.2012.64.i
12. Schneier, B.M.: Cryptanalysis of microsoft's point-to-point tunneling protocol (PPTP). In: Proceedings of the 5th ACM Conference on Communications and Computer Security, pp. 132–141. ACM Press (1998)
13. ENISA: Measurement Frameworks and Metrics for Resilient Networks and Services: Challenges and Recommendation. In: European Network and Information Security Agency (ENISA) (2010)

14. Monakhov, Y.M., Monakhov, M.Y, Luchinkin, S.D., Kuznetsova, A.P., Monakhova, M.M.: Availability as a metric for region-scale telecommunication designs. In: 10th IEEE International Conference on Intelligent Data Acquisition and Advanced Computing Systems: Technology and Applications (IDAACS), pp. 775–779. Metz, France (2019). https://doi.org/10.1109/idaacs.2019.8924390
15. Narayan, S., Williams, CJ., Hart, D.K., Qualtrough, M.W.: Network performance comparison of VPN protocols on wired and wireless networks. In: 2015 International Conference on Computer Communication and Informatics (ICCCI), pp. 1–7. Coimbatore (2015). https://doi.org/10.1109/iccci.2015.7218077
16. Kompella, S., et al.: Layer 2 virtual private networks using BGP for auto-discovery and signaling. In: IETF RFC, May 2012, https://tools.ietf.org/html/rfc6624
17. Metz, C.: The latest in VPNs, part II. In: IEEE Internet Computing, vol. 8, no. 3, pp. 60–65, May-June 2004, https://doi.org/10.1109/mic.2004.1297275 (2004)
18. Kent, S., Seo, K.: Security Architecture for the Internet Protocol (No. 4301). In: IETF RFC, December 2005. http://www.ietf.org/rfc/rfc430l.txt
19. Hauser, F., Häberle, M., Schmidt, M., Menth, M.: P4-IPsec: Site-to-Site and Host-to-Site VPN with IPsec in P4-Based SDN. In: IEEE Access (2020). https://doi.org/10.1109/access.2020.3012738
20. Qu, J., Li, T., Dang, F.: Performance evaluation and analysis of OpenVPN on Android. In: Fourth International Conference on Computational and Information Sciences, pp. 1088–1109. Chongqing (2012). https://doi.org/10.1109/iccis.2012.203
21. Kivinen, S., et al.: Negotiation of NAT-Traversal in the IKE(No. 3947). In: IETF RFC, January 2005 https://tools.ietf.org/html/rfc3947
22. Soucy, R.: Network Router Performance Testing How-To. http://soucy.org/vyos/NetworkPerformanceTesting.pdf. Accessed 08 Feb 2020
23. Vyos Homepage. https://www.vyos.io/. Accessed 13 July 2020

Routing and Transport

Multi-level Hierarchical Controller Placement in Software Defined Networking

Kurdman Abdulrahman Rasol and Jordi Domingo-Pascual(✉)

Department of Computer Architecture,
Universitat Politècnica de Catalunya (UPC), Barcelona, Catalonia, Spain
kurdman.rasol@upc.edu, jordi.domingo@ac.upc.edu

Abstract. Software Defined Networking (SDN) is a widely used network architecture. It separates the controller logic (or control plane) from forwarding plane (or data plane) to manage the whole network and it enables the network scalability and programmability. One of the most significant challenges in Software Defined Networking (SDN) is the Controller Placement Problem (CPP), which tries to specify the minimum number of controllers and their optimal location.

In our study, we extend the methodology based on K-means and K-center algorithms to solve the Controller Placement problem (CPP) into a Multi-level Hierarchical Controller Placement Problem (HCPP), where the Super Controller (SC) is in the top-level, some Master Controllers (MCs) are in the intermediate level and the Domain Controllers (DCs) are at the bottom level. The optimization metric is the latency between controller and switches assigned to it and the latency for controller to controller communication.

The proposed architecture and methodology is applied on Western European NRENs topology taken from Internet Topology Zoo. The entire network topology is divided into small scale networks (Clusters) and for each cluster, the optimal number of controllers (Domain Controllers) and their placement is found. A second optimization identifies the optimal number of Master Controllers and their optimal placement.

The results validate the methodology and show its feasibility on large networks and different domains. A useful use case may be the deployment of hierarchical levels of controllers for the enforcement of very precise routing policies through different domains.

Keywords: Software Defined Network · Controllers placement · Latency optimization · Hierarchical Control Plane · Multistage optimization

1 Introduction

The key idea of Software Defined Networking (SDN) is separation the forwarding plane (or data plane) and the control plane to manage network performance. SDN feasibility is supported by a logically centralized control plane [1,2].

B. Ghita and S. Shiaeles (Eds.): INC 2020, LNNS 180, pp. 131–145, 2021.
https://doi.org/10.1007/978-3-030-64758-2_10

The main focus of this paper is to minimize the control plane latency. Control latency comprises transmission delay, propagation delay, switch queuing latency and controller processing latency [3,4]. The transmission delay at the switch and controller refers to the ratio of packet size and the data rate of a link. The propagation delay depends on the distance between switches and controllers. The queuing latency is mainly determined by congestion in the path between source and destination. The controller processing latency is principally influenced by the load of the controller. In our current research, the metric of the optimization study is based on propagation delay and referred as latency.

The placement of the controller is a key design choice of the SDN control plane. The controller placement problem (CPP) is a non-deterministic polynomial NP-hard problem and is similar to the facility location problem [5]. Heller et al. [6] had proposed the principal objective to solve the Controller Placement Problem (CPP) by minimizing the propagation delay (latency) to compute the optimum location of the controllers deployed and the minimum number of controllers to be placed. The authors conducted experiments on the Internet2 OS3E network topology [7]. K-means or K-median is defined for minimizing average propagation delay [8], while K-center is for minimizing the largest distance latency between controllers and their associated switches [9]. For finding the optimum controller positions of the networks, the following questions must be solved:

1. What should be the minimum number of the controller to be placed in a network?
2. Where to place the controllers?

In this paper, we consider the scenario of Uncapacitated Controller Placement Problem (UCPP) [10]. This means that the capacity is not considered as a constraint and the controller is located with a switch [11].

Deploying a single controller is inefficient to manage a large network. Placing multiple controllers is still an outstanding challenge. For large size networks, multiple controllers are most efficient to handle the control plane traffic in order to improve the scalability and latency of the network [12]. Kuang et al. [13] proposed an algorithm based on a hierarchical K-means algorithm to solve a controller placement problem. The objective of the study is to minimize the maximum latency between the controller and its associated switches and to balance the load of each controller. The main contribution of the work is: a) the large network is divided into several single controller domains. b) In the process of network partitioning, not only latency between the switches and controller is taking into account but also the load balancing. The experimental results illustrate the partitioning used K*-mean algorithm it is proven to be more balanced than the optimized K-means algorithm but the propagation latency is larger than the optimized K-means algorithm by Wang et al. [8].

For very large networks composed of several Autonomous Systems (AS), we explore the feasibility of a Multi-level Hierarchical Control Plane. The multi-level hierarchical controller plane approach is efficient because the controller-controller

Symbol	Description
$G(V, E, L)$	Physical Network
V	Set of switches in the network (nodes)
E	Set of physical links between switches (edges)
L	Set of switch locations
C	Set of Controllers, where ($c \subset V$)
(i, j)	The link between node i and node j
$d(v, c)$	The distance between switch $v \in V$ and controller $c \in C$
n	Total number of switches or nodes in the network, where ($n = \|V\|$)
k	Total number of controllers to be installed in the network, where ($k \leq n$)
P	Set of all possible placement for k controllers
w_k	Indicates whether a controller is deployed at location k (= 1) or not (= 0)
$x_{i,k}$	Indicates whether switch i is connected to controller k (= 1) or not (= 0)

latencies must be considered along with the controller-switch latencies when evaluating the control plane reactivity perceived by the switches. The traffic on the controller-controller plane is crucial to achieving a consistent shared view of the network state that is the required condition to run network applications and the network state is stored in shared data structures. The response time of packets transmitted from the controller to each associated switches or from the switch to the controller is usually shorter than in the controller-controller traffic communication. In general, the controller-controller traffic is very complex because distributed controllers adopt coordination protocols and algorithms to synchronize their shared data structures to guarantee a consistent global network view and to enable a centralized view of the network state for the applications. The controller-switch traffic in control plane fundamentally depends on the network application running on the controller. As a result, the optimal controller placement in a given network should consider both kinds of latencies. On the other hand, our work concentrates on the controller placement problem by studying the impact of both latencies.

2 General Formulation of the Controller Placement Problem

The network is modelled as a graph $G = (V, E, L)$, where V represents the set of switches, E represents the set of links between switches and L represents the set of switch locations (longitude and latitude). To formulate the problem mathematically, $Lavg$ denotes the average latency. Let $d(v, c)$ be the distance (or edge weights) between switches $v \in V$ and the controllers $c \in C$. The model optimizes the latency between switches and controllers. It is also important to note that the number of controllers must be less than or equal to the number of switches $n = |V|$. The controller is deployed in the location of a switch. The number of controllers is fixed to k where $k \leq n$. The set of controllers to be installed is represented as $C = \{c_1, c_2, c_3, ..., c_k\}$. Therefore, the set of all possible placements (P) can be represented as $P = \{P_1, P_2, P_3, ..., P_m\}$, where m denotes the variations of n elements taken in groups of k without repetitions. The number of all possible placements (m) can be calculated as:

$$m = \frac{n!}{k!(n-k)!} \tag{1}$$

A reference study for the Controller Placement Problem in Software Defined Networking was done by Heller [6]. In the CPP, one of the most frequently utilized performance metrics is latency (or delay). Generally, latency describes the total time taken by a packet from the source node to the destination node. In this work, we take the latency as the distance between two nodes. In this paper, we study two types of optimization metrics which are average-case latency and worst-case latency [6].

2.1 Average-Case Latency

The average latency for the placement of the controller minimizes the average distances between the controller and switches assigned to the controller. This optimization problem is known as the minimum k-median problem or k-means problem [8]. The objective function can be defined as the follows:

$$L_{cs-avg}(P') = \frac{1}{n} \sum_{v \in V} \min_{(c \in P')} d(v, c) \tag{2}$$

Where $d(v, c)$ is the distance between switch $v \in V$ associated with controller c. The objective is to find the optimal placement of controller P' from the set of all possible controller placements where $|P'| = k$ and the $L_{cs-avg}(P')$ is the minimum.

2.2 Worst-Case Latency

The worst-case latency minimizes the worst (or maximum latency) between a switch $v \in V$ and its controller $c \in C$. This optimization problem is known as minimum k-center problem [9]. The k-center problem is a location analysis problem related to the optimization problem in the area of operation research. The k-center scheme provides an optimal solution minimizing the longest distance among nodes. The objective function is defined as follows:

$$L_{cs-worst}(P') = \max_{v \in V} \min_{c \in P'} d(v, c) \tag{3}$$

The optimization result comes from minimizing the previous optimization function. The goal of the optimization is to find the optimal minimum-latency placements P' from the set of all possible placements P.

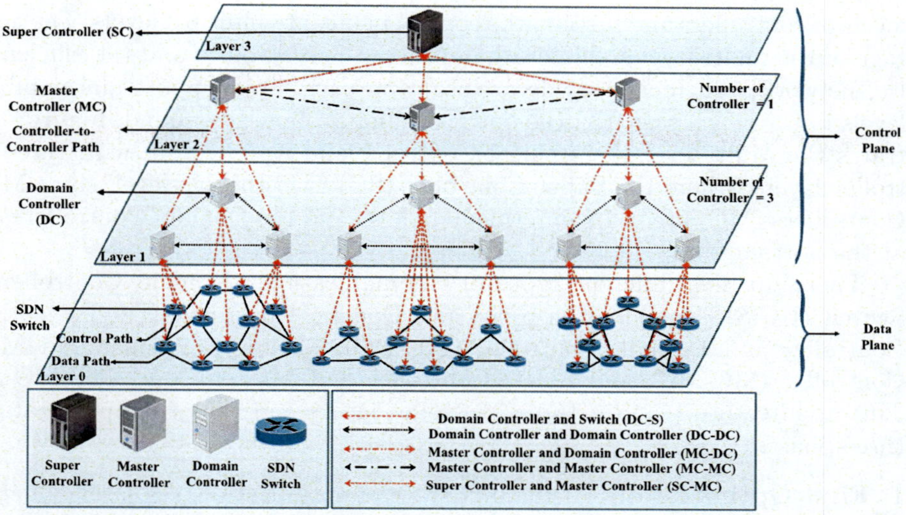

Fig. 1. Multi-level Hierarchical Control Plane Architecture

2.3 Formulation Constraints

The mathematical constraints are computed as follows:

$$\forall i \in V : \sum_{k \in P} x_{i,k} = 1 \tag{4}$$

$$\sum_{k \in P} w_k = k \tag{5}$$

$$\forall i \in V, \forall k \in P : x_{i,k} \leq w_k \tag{6}$$

$$\forall k \in P : w_k \in \{0, 1\} \tag{7}$$

$$\forall i \in V, \forall k \in P : x_{i,k} \in \{0, 1\} \tag{8}$$

Equation (4) ensures that each switch is associated with exactly one controller. Equation (5) restricts that the number of controller deployed is exactly to k. (6) signifies that switch i is located to controller k if controller k deployed onto switch i. Constraints (7) and (8) specify that the variables $x_{i,k}$ and w_k are binary (equal to 0 or 1).

3 Proposed Architecture and Methodology

The methodology proposed is named Multi-level Hierarchical Controller Placement Problem (HCPP). In the hierarchical architecture, the controller at the higher-level manages the controllers at the lower-level. There are at least two levels of controllers and it splits the control plane into multiple levels. The advantage of the multi-level architecture improves the scalability and the efficiency of the network [14]. Ideally, a hierarchical approach may support any number of levels but for each scenario, the optimal number should be found. Figure 1 illustrates the Multi-level Hierarchical Control Plane Architecture with three controller layers, where the Super Controller (SC) is in the top-level, some Master Controllers (MC) are in the intermediate level and the Domain Controllers (DC) at the bottom level.

There are five different types of communications: Domain Controller and Switch (DC-S), Domain Controller and Domain Controller (DC-DC), Master Controller and Domain Controller (MC-DC), Master Controller and Master Controller (MC-MC) and Super Controller and Master Controller (SC-MC). This architecture requires three controller layers and thus the optimization of three-phases:

1. **First Optimization:** At the first stage of optimization, the latency between switches and Domain Controllers (DCs) is minimized. All Domain Controllers (DCs) are at the one level and responsible to manage switches in the domain.
2. **Second Optimization:** The goal of the second optimization is to identify the number and placement of Master Controllers (MCs) where the Domain Controllers (DCs) act as switches (controlled devices). The topology for connecting Domain Controllers (DCs) is built as a virtual network (VN) based on the physical underlying topology and computing the shortest path between pairs of Domain Controllers (DCs).
   Dijkstra Algorithm is implemented on the top physical underlying topology to determine the shortest path for all pairs of Domain Controllers (DCs).
3. **Third Optimization:** In the upper-level, a super controller (SC) acts as a global controller which is logically centralized control plane. Again, a virtual topology connecting all Master Controllers (MCs) is built based on the physical underlying topology. Then, using *Dijkstra Algorithm* to compute the shortest path between all pairs of Master Controllers (MCs) to find the placement of the Super Controller (SC).

4 Evaluation and Results

4.1 Western European NRENs Topology

The proposed method to solve the Hierarchical Controller Placement Problem (HCPP) is applied on a set of networks of Western European National Research and Education Network NRENs (278 routers) shown in Fig. 2 (*left*). The data about Western European NREN is obtained from Internet Topology Zoo [15].

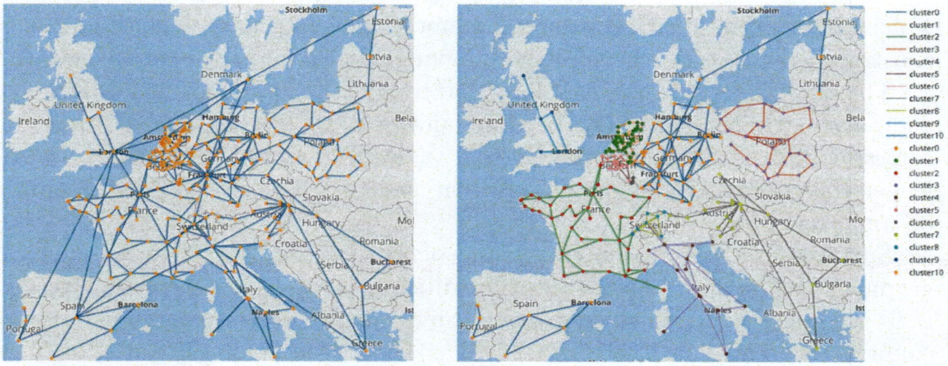

Fig. 2. The entire Western European NRENs Network Topology (left) and Western European NRENs Network Topology with Clusters (right)

Table 1. Cluster list

Country	Cluster	Nodes	Network
Germany	0	52	DFN
Denmark	0	1	GEANT
Estonia	0	1	GEANT
Latvia	0	1	GEANT
Lithuania	0	1	GEANT
Netherlands	1	51	SURFnet
France	2	39	RENATER
Poland	3	28	PSNC
Italy	4	20	GARR
Belgium	5	19	BELnet
Luxembourg	6	15	RESTENA
Austria	7	13	ACOnet
Slovenia	7	2	ARNES
Czech Republic	7	1	GEANT
Hungary	7	1	GEANT
Bulgaria	7	1	GEANT
Romania	7	1	GEANT
Greece	7	1	GEANT
Switzerland	8	12	SWITCH
United Kingdom	9	9	JANET
Spain	10	6	RedIris
Portugal	10	3	FCCN
Total	11 Cluster	278 Nodes	

We define each country as a cluster except the countries with less than six nodes, which are integrated with the nearest country. This foot tries to consider the real constraints derived from each network having its own management agency.

The data from Internet Topology Zoo presents seven nodes which are identified only by the name of the city or a country. We completed the database with corresponding coordinates of the city or the capital city of the country. In order to get the coordinates of the city, we checked with Google Maps and obtained its longitude and latitude.

Furthermore, we found out that some of the nodes are exactly in the same geographical location. The optimization results may select either of these locations.

The Western European network is divided into eleven clusters by using a country-based approach (see Fig. 2 (*right*). The reason for the partitioned network is to supervise the management of the large network. The clusters are listed on Table 1.

The linearization approach is implemented to transform the problem into mixed-integer programming (MIP) are solved optimally by using the well-know optimizer tool IBM ILOG CPLEX Optimization Studio V12.8.0 [16]. CPLEX is used to provide all optimal solutions and the model is programmed in Python.

The computing time varies from tenths of seconds to a few tens of seconds. The optimization running time to minimize the worst-case latency takes more time than the average-case latency but this is not relevant as it is an offline optimization.

4.2 First Optimization: Clustering-Based

The Latency Between Domain Controller (DC) and Switches. Figure 3 shows the Cumulative Distribution Function (CDF) for all possible combination placements of the controllers for cluster 1 by utilising a brute-force approach. The best placement corresponds to the lower latency for each value of k. We use cluster 1 (Netherlands) as an example. The optimum solution when the number of Domain Controllers (DCs) is one (k = 1) given an average latency of (L_{cs-avg} = 0.3363 ms) and the optimal controller location is in Utrecht. For the worst-case latency is ($L_{cs-worst}$ = 0.6890 ms) and the placement is in Amsterdam.

The latency between Domain Controller (DC) and their associated switches decreases gradually as more controllers added in the network.

Table 2 summarize the optimum latency and the locations of controllers outcomes of the optimization model for each cluster when the number of Domain Controllers (DCs) is varied from one to three. Running the optimization model is defined in Sect. 2.

Benefit to Cost Ratio. The goal of this metric is to decide if more controllers should be deployed. Adding a new controller reduces latency but adds cost. The Benefit to Cost Ratio is defined as:

$$(lat_1/lat_k)/k \tag{9}$$

Table 2. Optimal domain controller location for average-case latency and worst-case latency

Clusters	K = 1	K = 2	K = 3
Average-case latency			
C0	Frankfurt 1.1704 ms	Frankfurt, Potsdam 0.8832 ms	Frankfurt, Potsdam, Riga 0.6758 ms
C1	Utrecht 0.3363 ms	Hoogeveen, Utrecht 0.2531 ms	Hoogeveen, Utrecht, Breda 0.2044 ms
C2	Paris 1.4935 ms	Nantes, Lyon 1.0797 ms	Paris, Nantes, Lyon 0.7831 ms
C3	Lodz 1.0473 ms	Gdansk, Krakow 0.7796 ms	Gdansk, Radom, Gliwice 0.5990 ms
C4	Bologna 1.1376 ms	Cagliari, Milan 0.6373 ms	Rome, Milan, Cagliari 0.5202 ms
C5	Evere 0.2570 ms	Evere, Louvain-la-Neuve 0.2006 ms	Evere, Louvain-la-Neuve, Liege 0.1677 ms
C6	Kirchberg 0.0273 ms	Kirchberg, Ettelbruck 0.0171 ms	Kirchberg, Esch-sur-Alzette, Diekirch 0.0125 ms
C7	Vienna 1.0812 ms	Vienna, Sofia 0.7400 ms	Vienna, Sofia, Dornbirn 0.6315 ms
C8	Lausanne 0.4217 ms	Lausanne, Zurich 0.3013 ms	Lausanne, Zurich, Kreuzlingen 0.1940
C9	London 0.5830 ms	London, Warrington 0.2893 ms	London, Warrington, Glasgow 0.1732 ms
C10	Madrid 1.3200 ms	Madrid, Coimbra 0.7059 ms	Madrid, Coimbra, Barcelona 0.5427 ms
Worst-case latency			
C0	Copenhagen 4.4923 ms	Frankfurt, Riga 2.2377 ms	Copenhagen, Hannover, Riga 1.7625 ms
C1	Amsterdam 0.6890 ms	Zwolle, Eindhoven 0.5004 ms	Nijmegen, Leeuwarden, Tilburg 0.4164 ms
C2	Limoges 3.0267 ms	Rouen, Marseille 2.0376 ms	Paris, Nantes, Cadarache 1.6956 ms
C3	Lodz 1.8986 ms	Gdansk, Gliwice 1.3690 ms	Gdansk, Lodz, Bielsko-Biala 1.1531 ms
C4	Rome 2.3001 ms	Rome, Milan 1.7877 ms	Cagliari, Rome, Milan 1.2561 ms
C5	Vilvoorde 0.6553 ms	Antwerpen, Namur 0.4061 ms	Antwerpen, Namur, Evere 0.3754 ms
C6	Kirchberg 0.0904 ms	Hollerich, Diekirch 0.0526 ms	Diekirch, Esch-sur-Alzette, Limpertsberg 0.0307 ms
C7	Budapest 3.8529 ms	Vienna, Sofia 1.7499 ms	Vienna, Bucureşti, Athina 1.7150 ms
C8	Lausanne 0.7703 ms	Bern, Kreuzlingen 0.5063 ms	Bern, Kreuzlingen, Manno 0.4839 ms
C9	Warrington 1.1791 ms	Warrington, Telehouse 0.9859 ms	Glasgow, Warrington, Reading 0.3736 ms
C10	Madrid 2.6199 ms	Madrid, Lisboa 1.6848 ms	Lisboa, Sevilla , Valencia 1.0099 ms

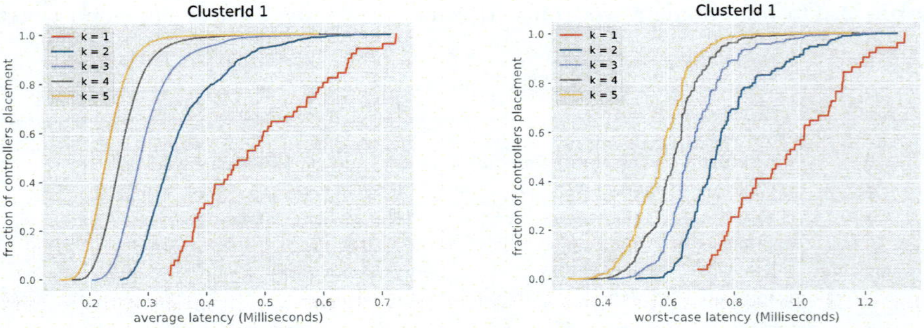

Fig. 3. Optimal latency CDFs for all possible controller combination placements for average-case latency (left) and worst-case latency (right)

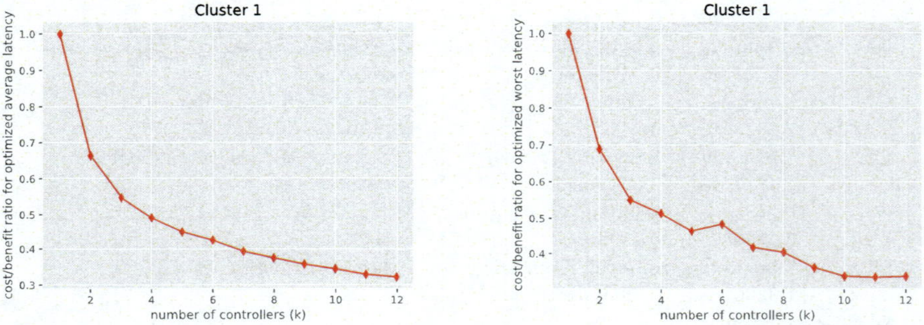

Fig. 4. Benefit to Cost Ratio for average-case latency (left) and worst-case latency (right)

Where lat_1 represents the optimal latency with a single controller, lat_k represents the latency with k controllers. The reason to divide by k to take into account in the cost of adding controllers [6].

Figure 4 shows the Benefit to Cost Ratio of deploying Domain Controllers (DCs) when the number of controllers is increasing from 1 to 12.

In this paper, we define a rule to make an easier way to determine how many controllers we need. The minimum number of controllers required corresponds to the latency $(lat1)/2$. Looking for a reduction of the latency when k = 1 to a half, we conclude that three to four controllers are suitable to optimize the average latency and four to five controllers for worst-case latency. We found out that the number of controllers is similar for the other clusters.

4.3 Second Optimization: Master Controllers (MCs) Placement

In the second stage of optimization, we generate the topology of a Virtual Network (VN) by using the physical underlying real topology interconnecting the Domain Controllers (DC) and switches. Dijkstra Algorithm is applied to compute distances and shortest paths between all pairs of Domain Controllers (DCs).

Figure 5 presents the latency between the Master Controller (MC) and Domain Controllers (DC). The optimal average latency is gradually decreasing by increasing the number of Domain Controllers (DCs) Fig. 5 (*left*).

On the other hand, results in the worst scenario indicates the instability in some optimal solutions due to the limitation of the data links between two points in the real world topology despite the fact others are relatively similar to the average scenario Fig. 5 (*Right*).

The results of the second stage of optimization show that a single Master Controller (MC) provides the best performance on the Western European NRENs topology.

The solution of the second optimization gives us a single master controller and it is the optimal solution for the control plane. This means that we do not need the third optimization to apply for this particular topology.

The Master Controller (MC) is the main controller that manages Domain Controllers (DCs) in different countries, while the Domain Controller (DC) manages switches in cities in the same country or neighbourhood countries.

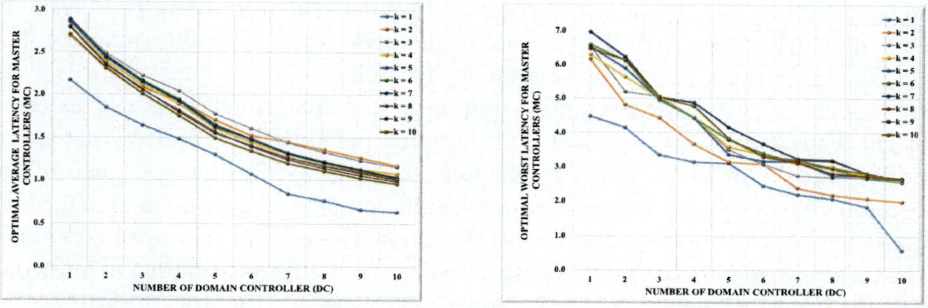

Fig. 5. The optimal solution for Second Optimization for average-case latency (ms) (left) and worst-case latency (ms) (right)

Figure 6 shows the particular case where four Domain Controllers (DCs) per cluster are deployed. This means that the total number of Domain Controllers (DCs) is 44 for the second optimization. Then, we build a virtual network topology (VN) for the 44 Domain Controllers (DCs). Dijkstra algorithm is implemented on the top of the physical underlying topology to determine the shortest path between each pair of Domain controller (DC). Then, we run the optimization to find the best placement of the Master Controller (MC).

For the average-case latency, Fig. 6 (*left*) shows the Master Controller (MC) is located in Frankfurt city (Germany) and its central geographical location. In addition, another significant factor is the massive connectivity between Frankfurt city and other cities in the network. Figure 6 (*right*) illustrates the results of the second optimization for the worst-case scenario. The Master Controller (MC) placement is CERN (Switzerland). CERN is located at Geneva.

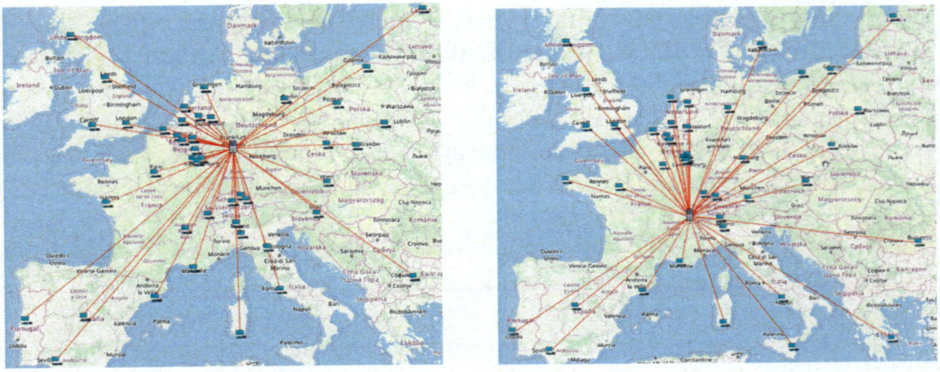

Fig. 6. The particular case for Master controller location for average-case latency (left) and worst-case latency (right)

4.4 Applicability Case Study

In this section, we compare the results obtained when applying the hierarchical controller placement against the case of a single level controller placement.

The main point is to demonstrate the feasibility of deploying multiple levels of controllers in very large networks composed of autonomous networks, as the case of the European NRENs. Each autonomous network has its own management policy and coordinates with its neighbours. Let's consider the case of inter-domain routing policies or end-to-end traffic engineering policies. Each network manager wants to have a close control on how the policies are applied within its network and, at the same time, wishes to disclose a limited amount of information and details to its neighbours. Domain Controllers are the appropriate approach for the application of intra-domain rules while Master Controllers may deal with the inter-domain rules.

For this comparison, we consider the average latency case. The whole set of NRENs is considered as a unique network and a single optimization step results in the optimal location of the controllers. Following the same optimization function and approach presented in Section II the average latency of one single controller is 2.1678 ms (this is lat1) and using the same rule to determine the best number of controllers needed (looking for lat1/2) results in six controllers (with an average latency of 1.0839 ms). Figure 7 (*left*) shows all results for optimal average latency of the Whole Western European NRENs without applying the Multi-level Hierarchical Control Plane methodology when the number of controllers is varied from 1 to 12. Figure 7 (*right*) presents the map with the location of the six controllers [(Vienna, Austria), (Amsterdam, Netherlands), (Frankfurt, Germany), (Milan, Italy), (Paris, France) and (Lodz, Poland)]. This result is referred as plain control plane.

The results obtained applying the hierarchical method gives that three Domain Controllers per cluster is the best option using the decision rule presented. Then, the second optimization results in a single Master Controller,

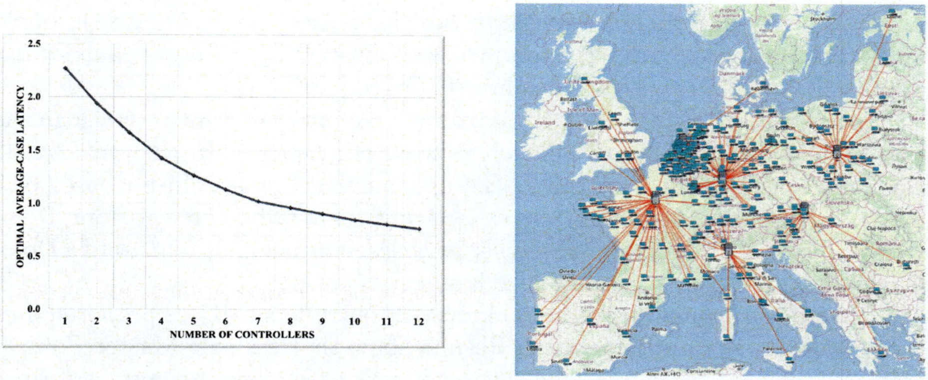

Fig. 7. The Optimal Average Latency of the Whole Western European NRENs for Plain Control Case (ms) (left) and The Whole Western European NRENs with Optimal Average Location when the number of controller is six (k = 6) for Plain Control Case (right)

which is placed in Frankfurt for the average latency case. This case is referred as Hierarchical Control Plane.

Comparing the results, for the plain control the average latency is 1.0839 ms. For the hierarchical plane, there is an average latency between the Domain Controller and Switch is about 0.4292 ms and the average latency between Master Controller and Domain Controller of 1.0839 ms. As most of the decisions may be made by the Domain Controllers the average latency is better with the hierarchical control but for some higher-level decisions, the average latency will be about the same than in the plain control approach. This is true only if all information from all the autonomous networks is shared and sent to the controllers in the plain control case.

From a practical point of view and real feasibility, we claim that the hierarchical control case allows detailed control within each domain and that the extra latency that may occur for some decisions made at the Master Controller will be compensated by the gain in management flexibility and autonomy.

5 Conclusion

Placement of controllers has received significant attention in recent years in the large scale networks. In this study, we minimize the latency between controllers and their associated switches.

This work presents a Multilevel Hierarchical Control Plane Architecture in SDN. The performance of this technique is evaluated with the real-world networks of the Western European NRENs). The topologies are taken from the Internet Topology Zoo collection. This is the first kind of recent CPP study is applied to the National Research and Education Network (NREN) in Europe.

The design methodology is performed from the bottom to the top because we consider the administrative boundaries; each NREN has its own management policy and this is the main constraints. For this reason, the real network topology (Western European NRENs) is divided into clusters by using a countries-based approach.The main contributions are the following.

As a result of the first stage optimization, we provide a rule to make an easier way to determine how many controllers we need. As a result, we found out that the number of Domain Controllers (DCs) is similar for each cluster based on the metrics; K-means and K-centre, respectively. Generally, three to four Domain Controllers (DCs) are required in case of average latency; or four to five Domain Controllers (DCs) in the worst-case latency scenario.

Second, the optimal placement of master controllers (MC) is proposed during the second phase of optimization. Results show that a single Master Controller (MC) is sufficient to manage the entire network to achieve the best performance in respect of this particular case study.

The overall contribution is the methodology presented to define a multi-level control plane as an iterative optimization problem. For this particular topology the results show that a third level is not needed. Other large topologies with different constraints may need a third level in the control plane.

All datasets and results will be made available for research purposes. Further research is progress to consider the reliability and capacity constraints of the Hierarchical Controller Placement Problem (HCPP).

Acknowledgements. This work was funded by the Spanish Ministry of Economy and Competitiveness under contract TEC2017-90034-C2-1-R (ALLIANCE).

References

1. Nunes, B.A.A., Mendonca, M., Nguyen, X.N., Obraczka, K., Turletti, T.: A survey of software-defined networking: past, present, and future of programmable networks. IEEE Commun. Surv. Tutor. **16**(3), 1617–1634 (2014)
2. Bannour, F., Souihi, S., Mellouk, A.: Distributed SDN control: survey, taxonomy, and challenges. IEEE Commun. Surv. Tutor. **20**(1), 333–354 (2018)
3. Kurose, J.F.: Computer Networking: A Top-Down Approach Featuring the Internet. Pearson, London (2005)
4. Wang, G., Zhao, Y., Huang, J., Wu, Y.: An effective approach to controller placement in software defined wide area networks. IEEE Trans. Netw. Serv. Manag. **15**(1), 344–355 (2018)
5. Arya, V., Garg, N., Khandekar, R., Meyerson, A., Munagala, K., Pandit, V.: Local search heuristics for k-median and facility location problems. In: Proceedings of the Annual ACM Symposium on Theory of Computing, vol. 33, no. 3, pp. 21–29 (2001)
6. Heller, B., Sherwood, R., McKeown, N.: The controller placement problem. In: ACM SIGCOMM Workshop on Hot Topics in Software Defined Networking, pp. 7–12 (2012)
7. Yeung, F.: Internet 2: scaling up the backbone for R&D. IEEE Internet Comput. **1**(2), 36–37 (1997)

8. Wang, G., Zhao, Y., Huang, J., Duan, Q., Li, J.: A K-means-based network partition algorithm for controller placement in software defined network. In: 2016 IEEE International Conference on Communications (ICC), Kuala Lumpur, pp. 1–6 (2016). https://doi.org/10.1109/ICC.2016.7511441
9. Sahoo, K.S., Sahoo, B., Dash, R., Tiwary, M.: Solving multi-controller placement problem in software defined network. In: 2016 International Conference on Information Technology (ICIT), Bhubaneswar, pp. 188–192 (2016). https://doi.org/10.1109/ICIT.2016.047
10. Singh, A.K., Srivastava, S.: A survey and classification of controller placement problem in SDN. Int. J. Netw. Manag. **28**(3), 1–25 (2018)
11. Hock, D., Hartmann, M., Gebert, S., Jarschel, M., Zinner, T., Tran-Gia, P.: Paretooptimal resilient controller placement in SDN-based core networks. In: International Teletraffic Congress, pp. 1–9 (2013)
12. Hu, T., Guo, Z., Yi, P., Baker, T., Lan, J.: Multi-controller based software-defined networking: a survey. IEEE Access **6**, 15980–15996 (2018). https://doi.org/10.1109/ACCESS.2018.2814738
13. Kuang, H., Qiu, Y., Li, R., Liu, X.: A hierarchical k-means algorithm for controller placement in SDN-based WAN architecture. In: 10th International Conference on Measuring Technology and Mechatronics Automation (2018)
14. Karakus, M., Durresi, A.: A survey: control plane scalability issues and approaches in software-defined networking (SDN). Comput. Netw. **112**, 279–293 (2017)
15. Knight, S., Nguyen, H.X., Falkner, N., Bowden, R., Roughan, M.: The internet topology zoo. IEEE J. Sel. Areas Commun. **29**(9), 1765–1775 (2011)
16. IBM ILOG. CPLEX Optimizer. http://www.01.ibm.com/software/commerce/optimization/cplex-optimizer (2012)

A Novel Congestion Avoidance Algorithm Using Two Routing Algorithms and Fast-Failover Group Table in SDN Networks

Seyed Hossein Mousavi Nejad[1](✉) and Mohammad Reza Majma[2]

[1] Department of IT Engineering (Computer Networks), Electronic Campus, Islamic Azad University, Tehran, Iran
h.mousavinejad@iauec.ac.ir

[2] Department of Computer Engineering, Pardis Branch, Islamic Azad University, Pardis, Iran
m_majma@pardisiau.ac.ir

Abstract. Today, the amount of data in the network is increasing quickly. As it does so, congestion in the network becomes more probable. Therefore, suitable policies must be used to control the congestion and guarantee quality of services. In this paper, we proposed an SDN based congestion avoidance algorithm. In our suggested method, we use an SDN controller to calculate link usage percent for all links in the network and predict congestion in every link. Additionally, we employ two types of Dijkstra algorithm; to calculate both the shortest path and the most secluded (the path with the lowest usage percentage) path between every node pairing in the network. Then, we store these two independent paths as two buckets of Fast-Failover group table in OF switches (these paths are recalculated and updated periodically based on updated traffic statistics of the network). After congestion recognition in an output link of a switch, we disable one of the input links of that switch. Consequently, the first bucket of that FF-group table entry will be ignored and second bucket will be employed to pass the traffic. Traffic will then be sent to its destination using the most secluded path instead of the shortest path until the usage percentage of congested link reduces to 50%. Finally, we developed our proposed algorithm in Python and used Mininet to emulate our network. We compare the algorithm's performance to one of the existing algorithms [1] of this type. Testing benchmarks showed 28% improvement in the average data transfer rate and 30% improvement in number of retransmitted packets in TCP mode. In UDP mode, we saw a 30% improvement in packet lost rate and a 24% improvement in average Jitter during data transfer.

Keywords: Congestion · SDN · SDN controller · Fast-failover group table · Open flow · Floodlight

1 Introduction

The growing amount of traffic in the network, coupled with the limited capacity of network links will strongly increase probability of link saturation and congestion occurrence in the network. This will lead to increased latency and packet loss – in other words reliability and QOS reduction.

B. Ghita and S. Shiaeles (Eds.): INC 2020, LNNS 180, pp. 146–158, 2021.
https://doi.org/10.1007/978-3-030-64758-2_11

In an SDN [2] environment, the central controller is capable of communicating with all switches, continuously monitoring their condition and gathering statistical data such as the volume of packets sent and received by them. Based on this data, it is possible to anticipate congestion before it occurs. To do this, we calculate the link utilization of each switch using the data provided by the controller. Therefore, congestion avoidance can be said consist of two phases; first we supervise all links continuously to find those with the higher usage percentage (Congestion Prediction). Then, we decrease the amount of data transferred through the congested link by employing an effective policy before the congestion can occur (Congestion Avoidance).

Congestion avoidance policies can be classified into several groups:

- Allocate resources surplus to requirements to some parts of the network [3]
- Avoid sending new data to the network for a period, in order to ease congestion [4]
- Accept new data and send it to the destination via one or several alternate paths to reduce primary path traffic [2, 5–9]
- Avoid congestion by controlling the data transfer rate [10].

In this paper, we used the third mechanism to avoid the congestion (after congestion prediction phase). Initially, more than one route is calculated and stored in switch's tables. After congestion detection, some of the link's traffic will be routed to this alternate route to decrease traffic in the link which is detected as a congested link and prevent it from becoming further congested.

In Sect. 2, we look at related works. In Sect. 3, we will describe our suggested algorithm in detail. In Sect. 4 we test our algorithm performance and finally in Sect. 5, we conclude our work.

2 Related Works

Lin et al. [11] employed the third method shown above (send new data to an alternate path) to resolve the congestion. To do so, they use Open Flow's Group Table capability. In their research, initially the controller calculates several routes between each two nodes using shortest path algorithm, then store each of them in an action bucket of the OF switch to redirect to an alternate route when failure or congestion occurs. In this method, all ports of switches are supervised periodically and an average transmit rate (ATR) is calculated for each port and compared to a threshold value (70% in this case) to detect congestion. After congestion detection, the switch replaces the congested route with another route which has the lowest ATR. When the ATR of the first port decreased to a value less than the threshold (70%), the primary route will be used again. This algorithm simulation showed that congestion time decreases about 47 to 72% less than normal situation. But in this method when a route selected as an alternative route for the congested one, the current situation of selected route does not get considered – meaning this route is not necessarily the most secluded route, and selection of it can precipitate congestion in this new route.

In another paper [1], a situation where all ports of switches are supervised periodically to detect congestion. When one port's usage percentage gets higher than 70, that switch will then be ignored, and a new route will be calculated with new topology which will replace current routes until usage percentage decreases to 50%. At this time, the deleted switch will be added to topology again. This algorithm is simple to develop. But its performance improvement is not significant. Also, in this algorithm, traffic is not considered when alternate routes calculated.

In a final paper [12], congestion detection is done by measuring the amount of data transferred in each port though a specific time period. Indeed, when the amount of data transferred from a link exceeds more than 70% of its capacity, it considered to be congested. At this point rerouting starts, but in this case all possible routes calculated again and the route with minimum load is selected by using data gathered from them. The positive aspect of this method is consideration of loads in every route.

3 Proposed Algorithm

In this algorithm, routing between hosts is done using two different methods:

1- Routing using the Dijkstra algorithm based on Shortest Path (SP): when link traffic is lower than our threshold (Normal Situation) the best path is the shortest one. So, in a normal situation we prefer to use this method as the default routing mechanism.
2- Routing using the Dijkstra algorithm based on The Most Secluded Path (TMSP): in this method, we receive statistics data of all ports from the Floodlight [13] controller using it's dedicated API and use these data to calculate occupation percentage and unused capacity of every link and use this as Weight parameter in Dijkstra algorithm to calculate the most secluded path between two nodes.

We choose to use a Floodlight controller as this controller supports both OF 1.3 and Fast-Failover Group Table (FFGT). In our algorithm, we first use these routing algorithms to calculate two different and independent routes between each pair of hosts. Then, these two routes are stored as two buckets in FFGT in the OF switches. When a packet enters a switch, if the first bucket output port (which conducts the packet to destination via the shortest path) is active, this path will be used. After congestion detection, we disable this port using the Mininet [14] API; so that the next packets enter that switch will be sent to the destination via the port of the second bucket instantly (via the most secluded path). Therefore, not only is this reaction done as fast as possible, but since the alternate path is selected based on present situation of the network traffic, it's not probable to face congestion in the new path.

Our proposed algorithm consists of two modules which run periodically:

1. Reroute Module (RM)
2. Congestion Detection Module (CDM).

This is flowchart of the first part of the algorithm: (see Fig. 1).

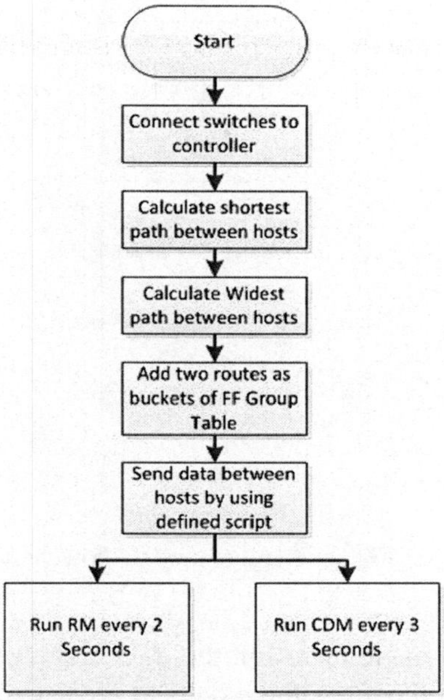

Fig. 1. Main flowchart

The lower timestamp between every run of these modules will lead to faster and better reaction in both congestion detection and path calculation. But we had two limitations to choose how often these modules run: first one is the limitation of floodlight statistic module which does not work properly in periods less than two seconds (as I mentioned at the end of section four). Also hardware resource limitations (such as CPU power) forced us to choose two seconds as frequency time for RM Module. In this module, at the beginning, statistics data was received from the controller. Then rerouting using TMSP is done based on the present traffic situation. New flow entries are being saved to tables with higher priority than previous flow entries. After calculating and saving the new entries, the older entries were removed from tables in order to reduce memory consumption. Therefore, we prevent high memory usage as while allowing the decision-making process to be done by the switch, as the new entries are added before deleting the older ones.

This is the flowchart of the RM Module: (see Fig. 2).

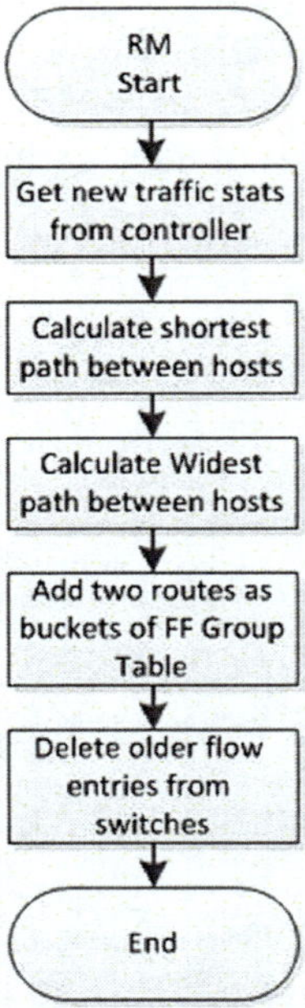

Fig. 2. Reroute module flowchart

Resource limitations made us to run the next module every three seconds to prevent CPU overload. This module's function is to predict congestion, do some effective reactions to prevent congestion by disabling suitable ports and reenable them when the congested link's load gets low enough. We tested several different values (60–85) as congestion threshold value and compare performance results of these values together and finally found out that the 80% is the most suitable and optimized value for it. Also we set the value of 50% as normal situation threshold.

For every port of the OF switches we get statistics about received traffic (RX) and transmitted traffic (TX). We use TX values of all ports to determine their traffic situation and detect congestion. So when the TX value of a port is higher than congestion threshold, in other words when output traffic of a switch becomes higher than congestion threshold, we check the RX traffic for all other ports of that switch to find the port which considerable amount of data enters the switch via that port (see Fig. 3).

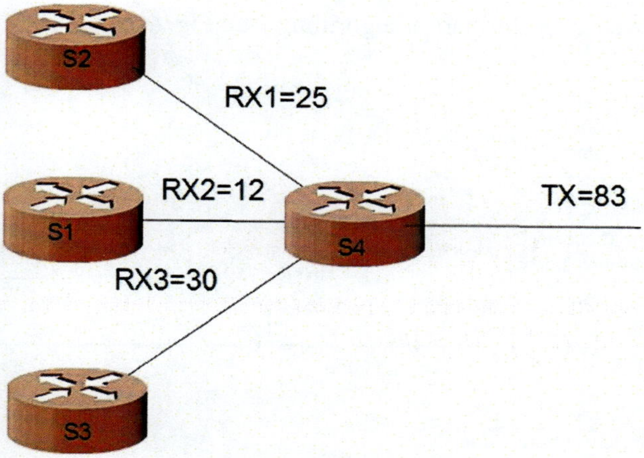

Fig. 3. Example of congestion reaction

Suppose that one port of switch 4 has a TX value of 83% and is therefore considered to be a congested link. We compare RX values of the other ports and choose the port that is connected to S3 as a high load input port. Then we disable the port of switch S3 which is connected to S4 using Mininet API and reduce the input traffic of S4.

At the beginning of this module, we check if there is any port which has been disabled before due to congestion detection. If any disabled port is found, that port's TX usage percentage is checked; if it is lower than 50 we enable that port because it has returned to a normal condition. Then, we receive all port's traffic statistics from controller, calculate TX usage percentage of each port, and compare them to threshold value to detect congested links. After this step, using the procedure I described before, we try to disable one of the input links to that switch with high load to decrease the traffic of congested link. After this process, FFGT mechanism conducts traffic from 1st bucket to 2nd bucket instantly and without any controller interference.

Now after disabling a port, we get new traffic statistics of congested link from the controller to make sure disabled link has significant effect on congested links traffic. If this effect was not considerable, we enable that link again and try to disable another input link with high load of data and check that effect on congested links traffic. We do this process for three times as needed to disable a suitable input link. But in our tests, the first deleted link usually has the most significant effect and there was no need to change it again.

This is flowchart of this part of algorithm: (see Fig. 4).

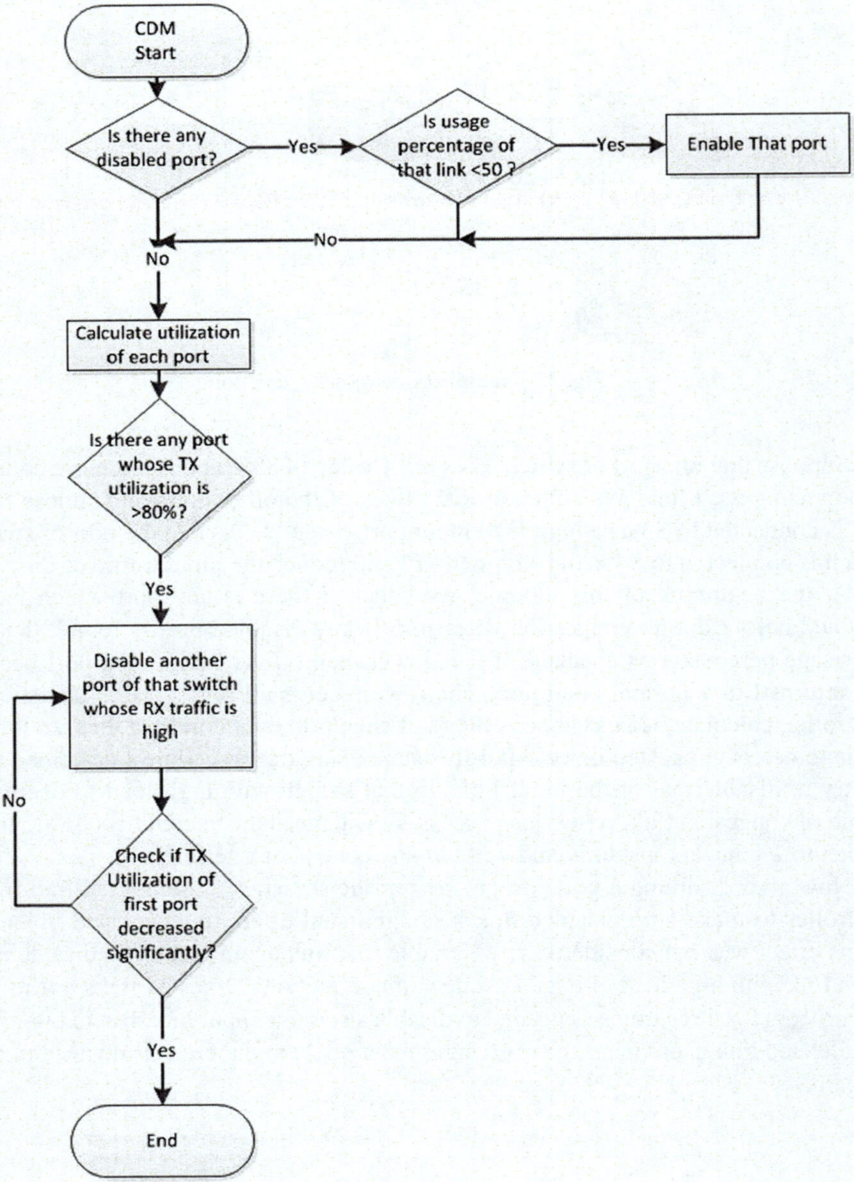

Fig. 4. Congestion detection module

4 Performance Evaluation

4.1 Test Environment

We run and benchmark our proposed algorithm on a VM[1] running Ubuntu 18.06. We used the latest version of Floodlight controller as the SDN controller. We developed the algorithm in python language using Pycharm software and used Mininet 2.2.1 to emulate our sample network topology. Finally, we have employed the 3rd version of iPerf software to send traffic between our nodes in the network.

4.2 Base Algorithm

The base algorithm [1] which the idea of our algorithm has been taken from is the following algorithm:

```
1. Start;
2. Connect switches to controller;
3. Set the shortest routing between hosts which need to
   send data;
4. Check the utilization of each port;
5. Is there any ports whose their utilization is more
   than threshold (70%)?
       a. Set New Route path between hosts without using
          that busy line;
       b. Is the utilization of busy path decreased under
          threshold (50%)?;
             i. Go To 3

       ELSE
             a. Keep using existing path;
          b. GO To 4;
   ELSE
       c. Keep using existing path;
       d. GO To 4;
```

4.3 Sample Topology

The sample topology we have used in our tests is shown in Fig. 5.

All switches are OVS switch and connected to floodlight controller.

To compare performance of two algorithms, we sent traffic from 4 clients to 4 servers simultaneously for one minute in our topology using an iperf script and saved iperf output results as text files to compare.

[1] Virtual Machine.

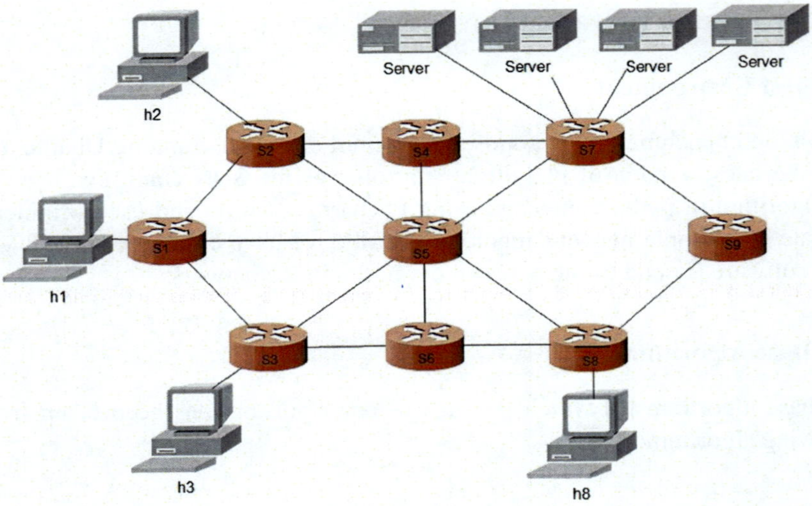

Fig. 5. Sample Topology

4.4 TCP Mode Performance

In the first test, we sent TCP traffic from clients to servers using iperf and calculate average bandwidth in both algorithms. We repeat this test 10 times and the proposed algorithm leads to 28% more bandwidth than the base one (see Fig. 6).

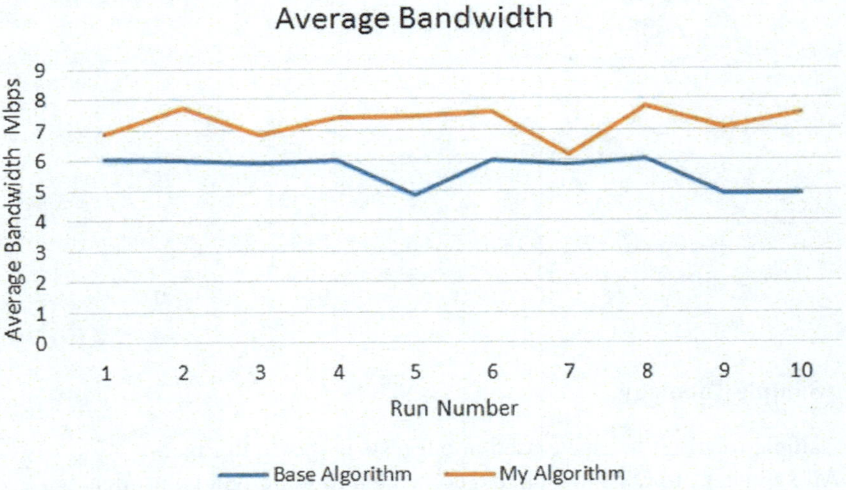

Fig. 6. Average bandwidth in TCP mode

In the next test, we compare number of ReT[2]s while sending traffic to servers using TCP protocol (see Fig. 7).

[2] Retransmit.

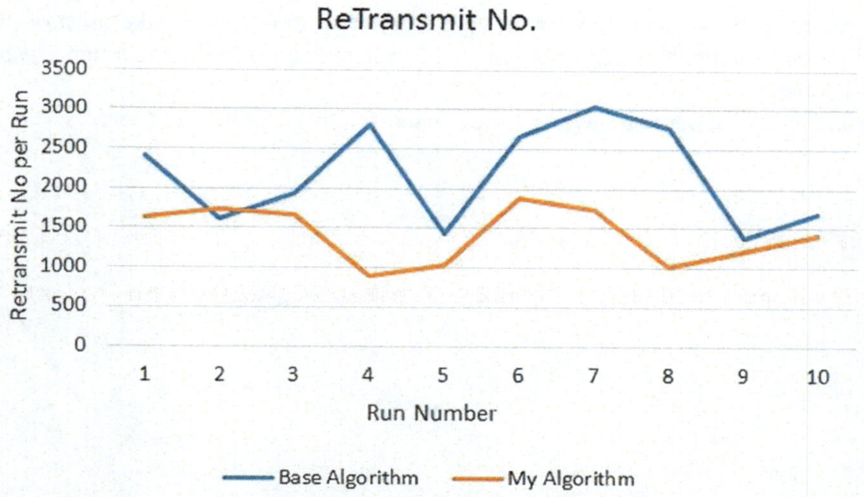

Fig. 7. Number of ReTs in TCP mode

This parameter is the total number of times TCP packets have been resent to the destination to ensure all sent packets have been received successfully at destination. A lower retransmit number means more successful and faster performance. In this category our algorithm showed an improvement of 30%.

4.5 UDP Mode Performance

In the next test, we sent UDP traffic to the servers and compared the results. Figure 8 shows packets lost (In Percent) using both algorithms.

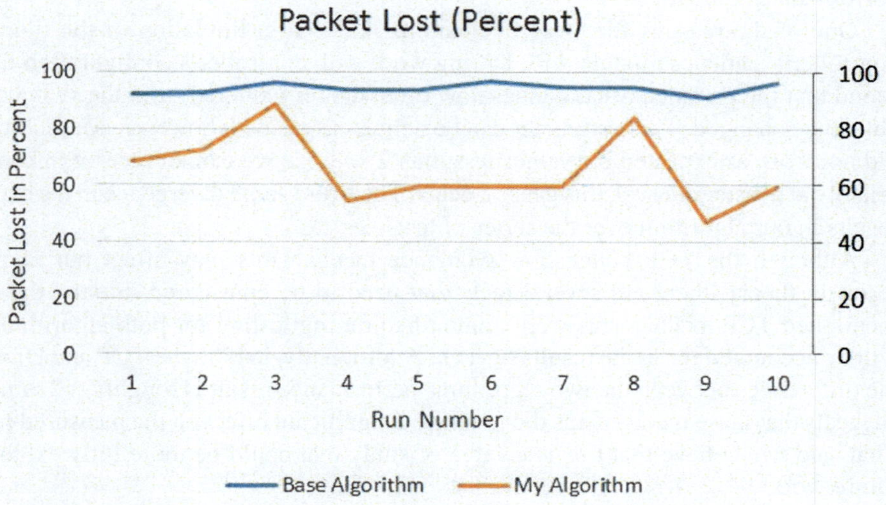

Fig. 8. Packet lost in percent in UDP mode

In this test, we sent traffic with high bandwidth in order to make most of network links approach the saturated situation. In this test, our algorithm's performance was about 30% lower.

In Fig. 9, we compare average jitter when sending UDP data to servers.

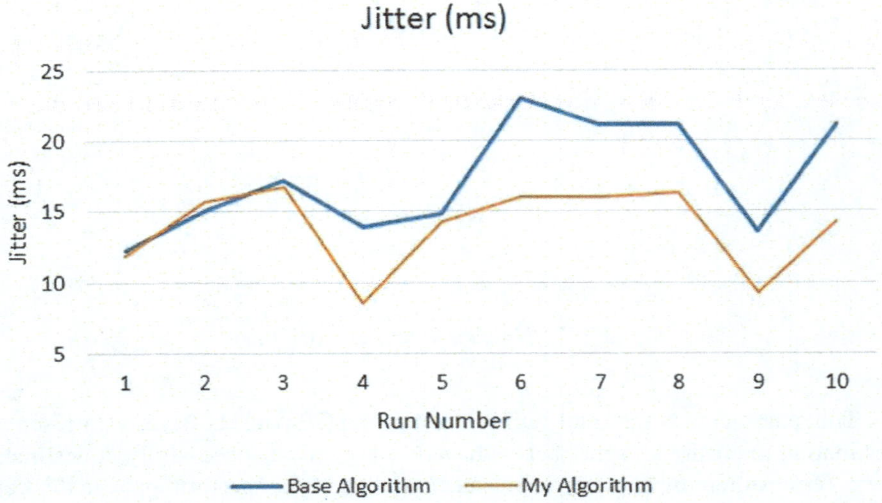

Fig. 9. Jitter in UDP mode

In terms of jitter, our algorithm's improvement was about 25%.

In all tests, some fluctuations were noticed in the results. Therefore, we ran every test 10 times in order to provide a representative sample of both our results, and the default algorithm. We then used an average of these tests in order to make our general performance comparison.

One of the reasons for these fluctuations is due to a limitation of the Floodlight controller's statistics module API. During work with controller's configuration file, we found that the parameter that defines how often data is gathered from the switches was able to be changed. This parameter can be defined as needed; however, we found that it did not work as expected for values less than 2 s. Since we cannot receive information reliably at a faster interval, this cause a delay, Therefore some divergence in the expected results in our algorithm over the series of tests.

Although the TCP congestion avoidance mechanisms may affect our algorithm result in theory, there are several facts that need to be considered. Firstly, the aforementioned TCP mechanisms were unmodified during testing for both algorithms, and will be accounted for in the results of each. Additionally, in both the TCP and UDP tests the differences between the two algorithms performance result is roughly the same; this suggests that these mechanisms did not have a significant effect on the measured results. That said, we believe that this is an area of study that could be more fully explored in future works.

In this paper, we compared our algorithm with another, which was the inspiration for our algorithm. In future studies, we would like to compare it against a broader selection of algorithms in this field.

Finally, hardware resource limitations led us to consider a simplified topology with a small number of hosts. In the future, we believe that this algorithm should be tested in both a larger and more complex computing environment.

5 Conclusions

Due to the large amount of data in today's networks, congestion of those networks is highly probable. In this paper, we proposed an algorithm which makes use of an SDN controller and OpenFlow capabilities in order to predict congestion in the network. We then reroute traffic to an alternate, secluded path in order to relieve high traffic routes, and prevent further congestion. We then compared our algorithm against another algorithm in this field; the benchmarks showed significant performance improvements in several different aspects.

References

1. Song, S., Lee, J., Son, K., Jung, H., Lee, J.: A congestion avoidance algorithm in SDN environment. In: 2016 International Conference on Information Networking (ICOIN), pp. 420–423. IEEE (2016). https://doi.org/10.1109/ICOIN.2016.7427148
2. SDN. https://www.opennetworking.org/sdn-definition/
3. Dikbiyik, F., Tornatore, M., Mukherjee, B.: Exploiting excess capacity for survivable traffic grooming in optical WDM backbone networks. GLOBECOM - IEEE Glob. Telecommun. Conf. **6**(2), 127–137 (2011). https://doi.org/10.1109/GLOCOM.2011.6133881
4. Domzał, J., Jajszczyk, A.: New congestion control mechanisms for flow-aware networks. In: IEEE International Conference on Communications, pp. 12–16 (2008). https://doi.org/10.1109/ICC.2008.11
5. Duli, Z., Wójcik, R., Dom, J.: Flow-Aware Multi-Topology Adaptive Routing. IEEE Commun. Lett. **18**(9), 1539–1542 (2014)
6. Nguyen, T.T., Kim, D.S.: Accumulative-load aware routing in software-defined networks. In: Proceeding - 2015 IEEE International Conference on Industrial Informatics, INDIN 2015, pp. 516–520 (2015). https://doi.org/10.1109/INDIN.2015.7281787
7. Kanagavelu, R., Lee, B.S., Miguel, R.F., The Dat, L.N., Mingjie, L.N.: Software defined network based adaptive routing for data replication in data centers. In: IEEE International Conference on Networks, ICON (2013). https://doi.org/10.1109/ICON.2013.6781967
8. Kabbani, A., Vamanan, B., Hasan, J., Duchene, F.: Flowbender: flow-level adaptive routing for improved latency and throughput in datacenter networks. In: CoNEXT 2014 - Proceedings of the 2014 Conference on Emerging Networking Experiments and Technologies, pp. 149–159 (2014). https://doi.org/10.1145/2674005.2674985
9. Kanagevlu, R., Aung, K.M.M.: SDN controlled local re-routing to reduce congestion in cloud data center. In: Proceedings - 2015 International Conference on Cloud Computing Research and Innovation, ICCCRI 2015, pp. 80–88 (2016). https://doi.org/10.1109/ICCCRI.2015.27
10. Gruen, J., Karl, M., Herfet, T.: Network supported congestion avoidance in software-defined networks. In: IEEE International Conference on Networks, ICON (2013). https://doi.org/10.1109/ICON.2013.6781970

11. Lin, Y.-D., Teng, H.-Y., Hsu, C.-R., Liao, C.-C., Lai, Y.-C.: Fast failover and switchover for link failures and congestion in software defined networks. In: 2016 IEEE International Conference on Communications (ICC), pp. 1–6. IEEE (2016). https://doi.org/10.1109/ICC.2016.7510886
12. Gholami, M., Akbari, B.: Congestion control in software defined data center networks through flow rerouting. In: 2015 23rd Iranian Conference on Electrical Engineering, pp. 654–657. IEEE (2015). https://doi.org/10.1109/IranianCEE.2015.7146295
13. Floodlight. https://floodlight.atlassian.net/wiki/spaces/floodlightcontroller/overview
14. Mininet. https://mininet.org

Impact of TCP Congestion Control Algorithms on HTTP/x Performance

Nicolás Illia[1(✉)] and Gabriel Tolosa[1,2]

[1] Departamento de Ciencias Básicas, Universidad Nacional de Luján, Luján, Argentina
nico@illia.dev, tolosoft@unlu.edu.ar

[2] CIDETIC, Universidad Nacional de Luján, Luján, Argentina

Abstract. Improving *web performance* is a significant concern for network engineers. Protocols at application and transport layers impose functional limitations, so different versions of them has been developed and deployed. HTTP/1.1 has been actively studied and revised to improve overall download speed, resulting in HTTP/2. This new version claims to make web downloads faster and more efficient. Complementarily, TCP's congestion control mechanism also imposes limits in data flows to prevent network congestion. However, the interaction between the two layers and their impact on the performance is not clear. In this work, we explore the performance differences in downloading a full web page using the two currently-used HTTP versions, according to the different congestion control algorithms implemented by TCP, controlling some critical network parameters. We show that there is not a unique combination that outperforms the remaining for all data flows, the different setups and networking scenarios.

Keywords: Congestion control · TCP · HTTP · Performance

1 Introduction

Over the past years, improving *web performance* has been a significant concern for network engineers and protocol designers. *Web performance* refers to page load speed from an arbitrary built web site, regardless of the underlying technologies that may exist in different parts of the network. Page load speed is highly related to the satisfactory user experience, which leads to better user engagement, retention and conversions [14]. Even more, page load time is one of the factors that Google takes into account to rank pages as a result of a search query [29].

For site developers, there is a growing trend to build faster web sites to match user expectations. This goal means taking advantage of different optimization techniques that lead to reductions on load time when a user accesses a given web page. The complete process considers all the elements that must be downloaded by the client application (i.e. the web browser) to render the full page correctly.

B. Ghita and S. Shiaeles (Eds.): INC 2020, LNNS 180, pp. 159–177, 2021.
https://doi.org/10.1007/978-3-030-64758-2_12

It is well-known that two critical components that shape the performance of network traffic are bandwidth and latency. However, other network parameters such as link capacity, router queues and error rates impose restrictions to data transfers.

Other key components that play specific roles and posses limitation are the protocols. It is evident that the overall performance is affected by the behaviour of the Hypertext Transfer Protocol (HTTP), including the existence (or not) of a security layer (SSL/TLS) or optimization techniques [19]. HTTP is one of the main building blocks of the Web and its use increases while more and more services are deployed to run into a browser window. The other key component of reliable Internet services is the TCP protocol [21]. Particularly, how it handles network congestion and limits the amount of data that is sent at each moment.

Another essential issue to consider is the size and complexity of web pages, which have been increased over time [11]. In the beginnings, pages were simple and contained a few external resources. The use of Javascript and Cascade Style Sheets (CSS) enabled many features to make pages look more beautiful and more functional, improving dynamism in client-side. One of the last style features is the possibility of making pages responsive to different screen sizes (i.e., computer, tablet, mobile phones) and resolutions which also increase the complexity of the HTML code and includes many external resources.

The size of web pages is also increasing for both desktop and mobile versions. According to HTTP Archive[1] the page size increased up to 327.6% and 1179,1% from 2010 to 2020 for desktop and mobile, reaching a median size of 1,999.9 KB and 1,852.1 KB, respectively.

The main effect of this growth in page sizes is an increase in traffic across the networks. Besides, new streaming services like Netflix and popular social networks like Instagram and Facebook push data transfers to continuously rise while requiring high download performance at the same time.

While data transfer technologies become faster and cheaper, physical distances posses limits to download speeds. The use of Content Delivery Networks such as Akamai or Cloudflare reduce distances between end users, and data providers help to mitigate performance bottlenecks. However, these distances cannot be reduced in all cases since CDN providers do not operate everywhere around the globe. Under these considerations, minimal improvements (i.e. a few milliseconds) in download speed may be significant for big Internet service providers. As an example, ten years ago Amazon estimated a cost of 1% in sales for each 100 ms of extra delay [28] while only two years ago Akamai estimated that the same delay might hurt their conversion rates by 7% [2] for customers completing online transactions. The same delay costs to Google 8 million searches which impact in the number of ads they display, thus impacting negatively in its business model [23].

It may be clear that the need for increased page download speed is not only a key factor regarding a satisfactory user experience but a requirement for companies to support current and future applications as well.

---

[1] https://httparchive.org/.

As we mention above, functional limitations are also posed by communication protocols. During the last decade, HTTP/1.1 has been actively studied and revised to improve overall client download speed. The main limitations are due to the fact that HTTP uses a single TCP connection to download each resource. Given that some web pages have dozen of links to external resources, the number of connections needed to download the full page generates overhead data (i.e. during connection opening and closing). Some of the HTTP/1.1 drawbacks are tackled by the development of HTTP/2, which is still being deployed.

This new version claims to make web downloads faster and more efficient by incorporating mechanisms to fix the head-of-line blocking issue and load page elements in parallel over a single TCP connection. It also compresses headers and introduces the server push feature [14]. However, it is unclear whether its adoption offers a real advantage or it is limited to particular cases. The reason lies in the fact that modern websites are already optimized to deal with HTTP/1.1 inefficiencies. Site designers employ server-side hacks such as spriting, inlining, and domain sharding [7].

The other protocol to take into consideration is TCP, specifically its congestion control mechanism [8]. This is an end-to-end, self-regulating traffic control whose primary goal is to prevent network saturation (i.e. incoming traffic in a router exceeds its outgoing bandwidth), thus limiting efficient data flows. From early implementations, TCP implements a mechanism to gradually increases the amount of data transmitted until the network's maximum capacity is reached [17]. Since then, different congestion control algorithms have been proposed and evaluated and adopted by separate TCP implementations [22] (i.e. TCP Tahoe, Reno, Vegas). The idea behind these improvements is to efficiently handle different networking environments such as high-speed or wireless networks [9].

However, there is no consensus about which congestion control algorithm to use and different operating system implementations use TCP variations that may handle outgoing traffic according to their strategy. For example, a modern Linux distribution as Ubuntu uses Cubic congestion control[2] while Windows family uses mainly Reno and CTCP [27].

It is essential to point out here that a new version of the HTTP protocol (HTTP/3) is still under developing, and its specification is in draft stage [6]. In the context of this work, the new HTTP/3 operates in a completely different way because the application protocol built up over QUIC [16] and, finally, over UDP [20] at the transport layer. In this case, congestion control mechanisms work across the entire connection, at QUIC level, so TCP is not involved. However, we propose this study as a baseline for a future performance comparison of HTTP/3 once it becomes deployed as a definitive standard.

1.1 Motivation and Goals

With the proliferation of web services and network traffic expansion, both HTTP efficiency and TCP congestion control capabilities are becoming increasingly

[2] See: */proc/sys/net/ipv4/tcp_congestion_control*.

important. Thus continuous improvement is becoming a research issue. However, different work tackled both topics in separate ways. By the one hand, congestion control algorithms and their performance have been extensively studied and compared under different networking scenarios [18], which generated recent new proposals. By the other hand, the development of HTTP/2, from its early stages based on the SPDY protocol [26], offered many insights about its performance under different web pages characteristics (i.e. diverse number of external resources and their sizes.) and web site optimization techniques used (i.e. domain sharding). HTTP/2 was revised in 2015, and the last update was released early in 2020, adding support for TLSv1.3 [5]. However, these reviews target mainly security concerns but do not improve some of HTTP/2's known deficiencies [12], in general, inherited from the SPDY protocol.

Under this scenario, we tackle some research issues that explore the performance differences on downloading a full web page using the two currently-used HTTP versions, according to the congestion control algorithm used by the TCP implementation at the transport layer. It is quite clear that it is not accurate enough to analyze the efficiency of an algorithm in isolation, and it is also essential its interaction and complement with other network mechanisms [18]. Specifically, we want to answer the following research questions:

- **RQ1:** Is there significant performance variability between the two HTTP versions under different TCP congestion control techniques (TCP-CC)?
- **RQ2:** How network parameters (error rate, latency and bandwidth) impact in the performance of different combinations of HTTP/TCP-CC?
- **RQ3:** Is it possible to establish recommendations about which combination of HTTP version, TCP Congestion Control algorithm and network characteristics benefit the download speed for different web pages?

Motivated by these uncertainties, in this work, we examine the performance of the diverse protocol combinations through controlling some important network parameters change such as bandwidth, delay and error rate. To the best of our knowledge, this is the first study that measures the HTTP/x download speed related to the congestion control strategy implemented by TCP and offers some insight about the different setups, networking scenarios and performance change thresholds.

The remainder of this paper is organised as follows: Sect. 2 briefly introduces some basic concepts and important work related to ours. Section 3 explains the proposed methodology. Section 4 shows the experiments and achieved results while Sect. 5.5 graphically summarises the main findings. Finally, we conclude in Sect. 6 and propose some future work.

2 Background and Related Work

As the applications over the Internet are expected to support more and better data flows, new strategies and algorithms are designed to prevent and manage the congestion in the network. TCP's Congestion Control is a critical mechanism that aims to keep the network stable while more bandwidth is allocated for outgoing segments within a connection.

Congestion control algorithms exploit the fact that TCP segments arrive at the receiver at a rate the network supports without generating a bottleneck. This rate is measured according to the received ACK segments, which enables TCP to estimate the current network status. In case of congestion, the ACK segment does not make to arrive on time, and TCP switches its mechanisms to a more conservative state. Otherwise, under a congestion-free scenario, TCP increases the sending rate. Different congestion control algorithms are developed and evaluated in the network research community [1], which are then adopted by various implementations of the main operating systems in their TCP/IP stacks. In general, they differ on how to measure the perceived congestion and increase the data sending rate. For example, TCP NewReno [4] or TCP CUBIC [15] use packet loss as the indication of network congestion.

There is a substantial body of work related to separate aspects of ours. Here, we briefly summarize some of them due to space constraints. Grieco and Mascolo [13] evaluate by simulation different congestion control algorithms in several aspects such as goodput, fairness and friendliness. Their study does not cover the use of the algorithms for specific applications. Similarly, Jaeger et al. [18] study a BBR implementation. They show that BBR has difficulty to reach a fairness equilibrium, thus suppressing other competing algorithms. In recent work, Turkovic et al. [24] study and evaluate the interactions among different congestion control algorithms. Their main result shows that fairness in resources claimed is not often achieved (mainly for flows sharing a link that have different RTTs). This finding means that some algorithms get improvements in performance only in specific situations.

In the case of HTTP/2, early analysis envisioned considerable performance benefits with regards to HTTP/1.1. However, as discussed by Akhshabi [3], the adoption of a new protocol depends on its benefits and its costs. Many of current websites have faced HTTP/1.1 inefficiencies by the use of hacks such as domain sharding and spriting, among others to reduce the page load time. This situation opens the question of whether and under what conditions HTTP/2 outperforms its predecessor. Some studies [10,25] show both improvement and degradation between them so this disagreement motivates further investigation.

But all these studies analyse the behaviour and performance of TCP and HTTP in separate ways. In this work, we investigate the interaction between transport and application layer protocols working together to decrease page load times. Given the previous findings, we go more in-depth on the understanding of the most competitive combinations for different data flows or web page layouts.

3 Methodology

The proposed methodology to tackle our research questions of analysing the target parameters is based on controlled real transfers of different web pages. We build a benchmarking scenario based on a Linux router, a web server and a web client, as shown in Fig. 1. Each machine has an Intel i5 (2.7 GHz) CPU with 8 GB of RAM, a 320 GB HDD and Gigabit Ethernet network cards. Both the client and server run the Ubuntu 17.10 Linux distribution, which is a common platform for running the mentioned applications. It also runs a Linux Kernel version 4.13 which has different TCP congestion control algorithms available. Moreover, the router is based on FreeBSD 11.1 because it provides the required bandwidth shaping and monitoring tools. To emulate the different scenarios, we use DummyNet[3] because it enables us to configure *pipes* with specific features (i.e. loss rate and bandwidth.) that manage packets I/O traffic. In the case of the web server, we run Nginx[4] because it supports the different HTTP versions to test, enabling TLSv1.2[5] where appropriate.

For each experiment, we capture all traffic using `tshark`[6] on the client-side while the server stores communication statistics. To this aim, we enable TCP Probe kernel module that allows us to check some essential parameters (i.e. `cwnd` values for each transmitted TCP segment, the `sshthresh` value, mainly.).

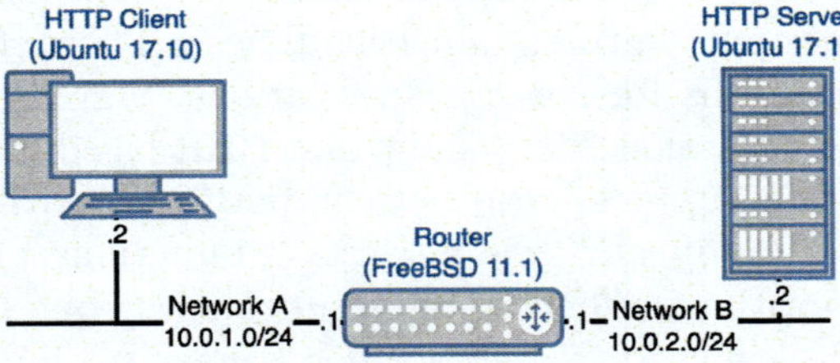

Fig. 1. Benchmarking scenario.

It is worth mentioning that any web server must be finely tuned to handle many concurrent connections. It also becomes necessary to adjust some parameters at the operating system level[7]. To isolate the target measures, we assume no bottlenecks at the software level, so we tune the web server according to general recommendations [14].

4 Experiments

4.1 Data

To test the configurations mentioned above under different networking scenarios, we use four datasets selected according to the following criteria: (1) a single file

[3] https://cs.baylor.edu/~donahoo/tools/dummy/.
[4] https://www.nginx.com/.
[5] We use a self-signed wildcard certificate for each emulated web site.
[6] https://www.wireshark.org/docs/man-pages/tshark.html.
[7] For example, the number of queued incoming connections, connection keep-alive time and window size. These parameters may be controlled using the `sysctl` command.

of 5 MB to test a continuous flow of bytes resembling, for example, an audio file. This data is identified as *continuous_flow*. (2) a common web page used in previous work to test HTTP/2 that is composed mainly by small flags of several countries [11]. This kind of page has only a few amounts of bytes, but it is composed of many resources, and it is intended to exploit HTTP/2 multiplexing. We refer to this dataset as *flags*. (3) a popular web site with the predominance of text. In this case, we use the main page of a Twitter user. (4) a popular web site with the predominance of multimedia content. Here, we use the main page of the logged user on Netflix. Table 1 shows the distribution of the total size among the different resources that form the last two datasets[8].

Table 1. Distribution of resources by type in Twitter and Netflix pages.

Type	Twitter			Netflix		
	Qnt	Size	%	Qnt	Size (bytes)	%
HTML	1	770,566	13.89	1	538,475	7.23
Scripts	6	2,565,307	46.26	1	530,785	7.13
Stylesheets	3	640,915	11.56	4	3,798,071	51.03
Images	32	1,506,471	27.16	139	2,405,770	32.32
Typefaces	1	25,092	0.45	1	73,572	0.99
Other	7	37,451	0.68	2	96,716	1.30
Total	50	5.545.802	100.00	148	7,443,389	100.00

Configurations: We test four TCP congestion control algorithms (CCA), namely: NewReno, Cubic, Westwood+ and BBR. Besides, we test three HTTP variants: (a) HTTP/1.1 (identified as **h1**). (b) HTTPS: In our case, HTTP/1.1 over TLSv1.2 (**h1s**). (c) HTTP/2 using also TLSv1.2 (**h2**). Then, we define the following network parameters[9]:

- $BANDWIDTH(bps) = \{1024, 2048, 5120, 10240, 20480\}$
- $DELAY(ms) = \{10, 25, 50, 125, 250, 500\}$
- $LOSS(\%) = \{0.00, 0.02, 0.05, 0.10, 0.20\}$

5 Results

The number of experiments to carry out according the different combinations of network parameters and protocol versions (Sect. 4) is large, so we decided to report here the a subset of the results that enable us to understand the interaction between the protocols. For each network parameter to test, we restrict the spectrum of values following these criteria:

[8] To control the number of bytes transferred we block some scripts that may download contents in running time. In the case of sites that implement the *infinite scrolling* technique to avoid pagination, we perform 5 scroll actions to complete the page.

[9] For each combination, we run 10 trials and averaged the results.

- For BW, we consider: Low (1 Mbps), Medium (5 Mbps) and High (20 Mbps).
- For LOSS, we select 2 cases: Moderate (2%) and High (10%).
- Finally, for DELAY, we select 2 cases[10]: Low (10 ms) and Medium (125 ms).

To get a clear insight about the interaction between the four TCP congestion control algorithms and the HTTP variants under different networking scenarios, we divide the results according to the dataset used for testing because these have different properties that impact on the overall behaviour. As we introduce in Sect. 1, the metric to optimize is the download speed of the full web page, including all its internal resources to allow the correct rendering, as presented to the user.

5.1 *continuous_flow* Dataset

The first set of experiments corresponds to the download of a single file (*continuous_flow*). In this case, we report results regarding HTTP/1.1 and HTTP/2 (h1s performs similarly to h1). Initially, we analyse the impact of the packet loss rate. As an overall observation, there are no apparent differences between h1 and h2, no matter the congestion control algorithm used. This result is expected because the transfer of only one resource does not take advantage of HTTP/2's features.

The differences appear among the CCA. As an example, Fig. 2 shows the results for 20 MB of bandwidth. HTTP/1.1 (dotted lines) and HTTP/2 (solid lines) perform similarly for the same underlying congestion control algorithm. When the link delay is low (10 ms) we do not observe significant differences for loss rate up to 10%, but then, these differences grow up to 53%. For a delay of 125 ms the performance of HTTP over *Cubic*, *Westwood* and *Reno* decreases linearly with the loss rate, while HTTP over *BBR* grows much more slowly. An unexpected result is the poor performance of *Cubic*, being the worst CCA that performs in this scenario. The other two bandwidth configurations show similar trends (we omit figures) with increased download times.

To deepen what was observed in the previous experiments, we analyse the impact of the network delay to download the same data. We consider the same configurations and report here the results for two packet loss scenario: moderate loss (2%) and high loss (10%). The overall picture is similar to the previous case: h1 and h2 perform in a similar way for the same underlying CCA. Figure 3 shows the results for 5 MB of bandwidth.

Although BBR is the most performing congestion control algorithm, there is an interesting observation among the others. In general, the second-best CCA is Westwood; however, when using 20 Mbps of bandwidth and the delay increases up to 500 ms, Cubic becomes better than Westwood (improvement up to 13%).

As a means to understand the different behaviour of BBR and the remaining CCA, we explore the size of `cwnd` while downloading the *continuous_flow*. Figure 4 shows the evolution of this parameter in time. It is possible to see that BRR has low variations of the congestion windows.

[10] We omit high delay links such us a satellites because these are out of this study.

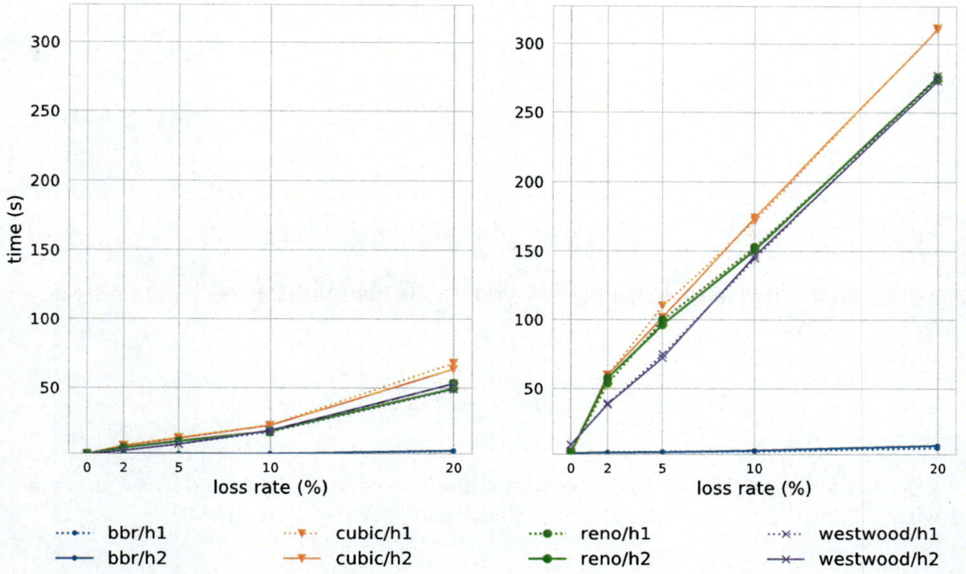

Fig. 2. Impact of loss rate on download time for the *continuous_flow* and 20 Mbps of bandwidth for a link-delay of 10 ms (left) and 125 ms (right).

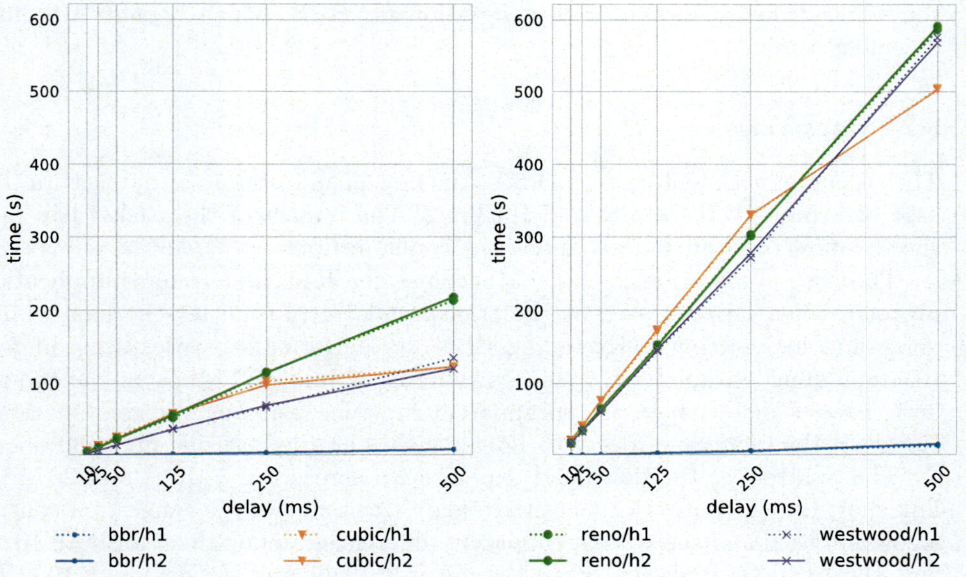

Fig. 3. Impact of link delay on download time for the *continuous_flow* dataset and 20 Mbps of bandwidth for a packet loss rate of 2% (left) and 10% (right).

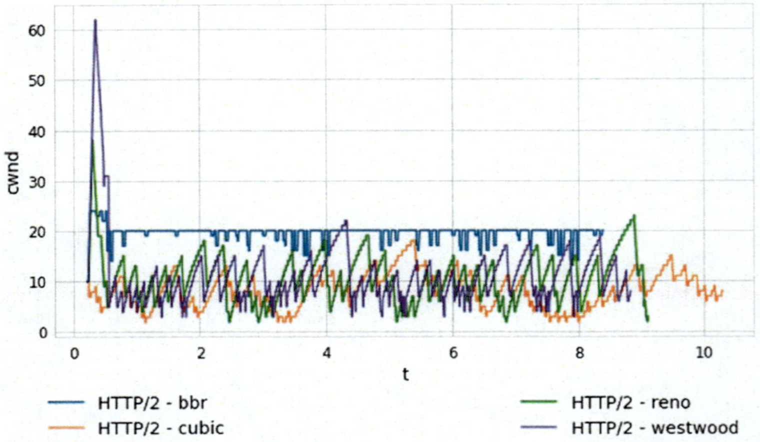

Fig. 4. Evolution of `cwnd` while downloading the *continuous_flow* dataset using a link with 2% of packet loss rate, 10 ms of delay and 5 Mbps of bandwidth.

The average value of `cwnd` is 19 for BBR, 8 for Cubic and 11 for both Reno and Westwood. Besides, standard deviation is 1.8, 3.1, 5.0 and 8.3, respectively. Although Reno or Westwood get higher `cwnd` values at some points in time, the whole transfer shows a lower variation for BBR, which benefits the final download time.

5.2 *flags* Dataset

The second experiment corresponds to the download of the *flags* dataset. In this case, we compare HTTPS[11] and HTTP/2. The impact of the packet loss rate shows interesting differences among the configurations.

The first observation is that h2 becomes the best performance application protocol when BBR is used at the transport level (no matter the bandwidth, delay and loss settings). Excluding BBR, its performance varies according to the remaining parameters. When considering a link of 1 Mbps of bandwidth and 10 ms of delay, there are no apparent differences among the combinations. However, the increase of delay to 125 ms makes h1 the best one up to 30%.

The analysis of the download experiments across the 5 Mbps and 20 Mbps link show that h1 outperforms (on average) h2 no matter the congestion control algorithm used and there are no apparent differences among them for both 10 ms and 125 ms links. However, when the delay is 10 ms and packet loss ≤10%, h2 outperform h1 using Reno or Westwood. Finally, with a link delay of 125 ms, the best performance is achieved using h1/Cubic.

[11] To allow a fair comparison due to it uses TLS in a similar way to HTTP/2.

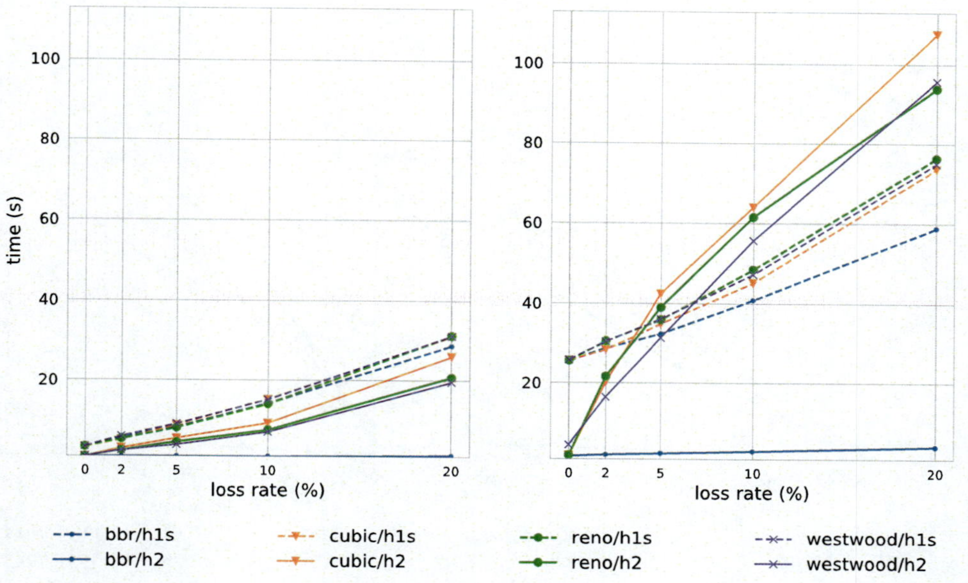

Fig. 5. Impact of loss rate on download time for the *flags* dataset and 20 Mbps of bandwidth for a link-delay of 10 ms (left) and 125 ms (right).

The analysis across different values of delay and packet loss is also interesting. In the case of a low delay (10 ms), h2 performs consistently better than h1s (Fig. 5, left) but there is a different behaviour when the link delay increases.

With a packet loss rate greater than 5% h1s becomes better than h2 when using Cubic, Reno or Westwood. That is, h2 is better than h1s when the congestion control algorithm is BBR. In all cases, BBR allows both h1s and h2 perform the best for 10 ms and 125 ms of link delay, respectively.

Then, we analyse the impact of delay for the two considered scenarios of packet loss. As in the previous case, the best performance application protocol changes according to the packet loss rate, except when BBR is used as congestion control. In this case, h2 is always the best option to download the flags dataset regardless of the link delay. Figure 6 shows an example for 20 Mbps of bandwidth (other bandwidth configurations show similar trends).

As a summary, we observe that in scenarios with low delay, h2 is always better than h1 and h1s, regardless of the underlying congestion control algorithm. Considering a link delay of 125 ms, h2 is better up to 2% of packet loss rate. Then, the performance of h2 depends on the congestion control algorithm. Up to 5% of packet loss rate, h2 performs better than h1s using Westwood or BBR. Above this value, the performance is determined by the use of BBR, although there is a difference of about 87% (on average) between h2 y h1s.

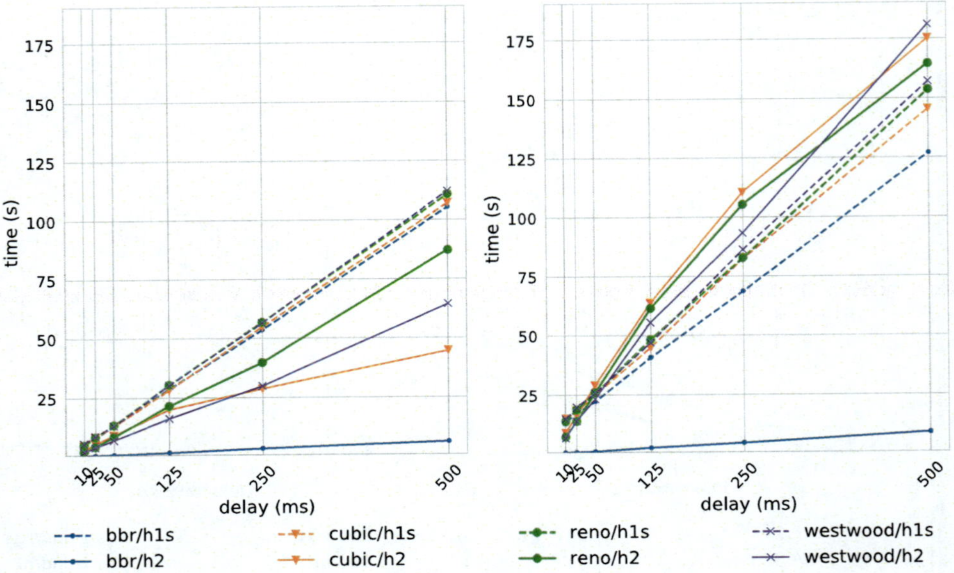

Fig. 6. Impact of delay for the *flags* dataset and 20 Mbps of bandwidth for a packet loss rate of 2% (left) and 10% (right).

5.3 *Twitter* Dataset

Following the same criteria as in previous experiments, we analyse the three different scenarios of link bandwidths. In the first case (1 Mbps, 10 ms), h1 is the best choice on average, but there are no apparent differences among the congestion control algorithms (less than 1% between BBR and Westwood). However, dissecting the results, it is possible to note that h2 performs the best for low packet loss, whereas h1 is the most efficient with high values of packet loss. Then, when considering a link of 125 ms of delay, h2 performs the worst no matter the congestion control algorithm used. The best combinations are h1/BBR and h1/Reno (+8.3%).

In the case of the higher bandwidth links, the situation is different. Here, h2/BBR is the most efficient combination. In the case of a 5 Mbps of bandwidth, its performance becomes −3% and −60% for 10 ms and 125 link delays, respectively. However, this situation reverses when excluding BBR and h1/Reno becomes the best performance one (close to h1/Westwood by around −1.0%).

For a 20 Mbps link, the differences are remarkable: h2/BBR is 80% faster than h1/Westwood (the second one). We also analyse the impact of the packet loss rate when downloading the *Twitter* dataset. The first observation is that h1s performs better than h2 when using Cubic, Reno or Westwood at the transport layer. However, this behaviour reverses when BBR is used as the congestion control algorithm (for both RTT scenarios, 10 and 125 ms). As an unexpected fact, Cubic performs the worst, mainly when h2 is used at the application layer. Figures 7 and 8 show the behaviour for a 20 Mbps link.

5.4 *Netflix* Dataset

The first analysis corresponds to link delay for both scenarios (10 and 125 ms). Similarly to the previous case, h1s performs better than h2 when using Cubic, Reno or Westwood at the transport layer but the most competitive combinations appear with BBR. In fact, h2 is the best performing application protocol no matter neither the RTT value nor the packet loss rate. Again, Cubic performs the worst for both h1s and h2 application protocols. Figure 9 shows the behaviour for a 20 Mbps link.

When analysing the behaviour according to the RTT dimension (Fig. 10), this looks similar. However, the four series for h1s perform rather close up to 125 ms of delay (less than 25% between ends values) with 2% of loss, but the h1s/BBR combination performs significantly better with 10% of packet loss.

The analysis in depth is divided according to the use (or not) of BBR as congestion control. For a 1024 Mbps link the best performing combination is h2/BBR (−7.4% over h2/Cubic) with low delay. When a delay of 125 ms h1/BBR is 25% better (on average) than the remaining combinations (it is not a clear second best). Excluding BBR, there is not a precise best combination due to all the remaining congestion control algorithms manage a limited channel (1 Mbps).

The scenario of 5 MB of bandwidth shows exciting pictures too. Using BBR and 10 ms of delay, there are no significant differences that establish a trend. However, with 125 ms of delay, the combination h2/BBR perform the best, outperforming h1/Westwood by around 58% (on average). Excluding BBR, the best combinations are h1/Cubic and h1/Westwood for 10 ms and 125 ms, respectively.

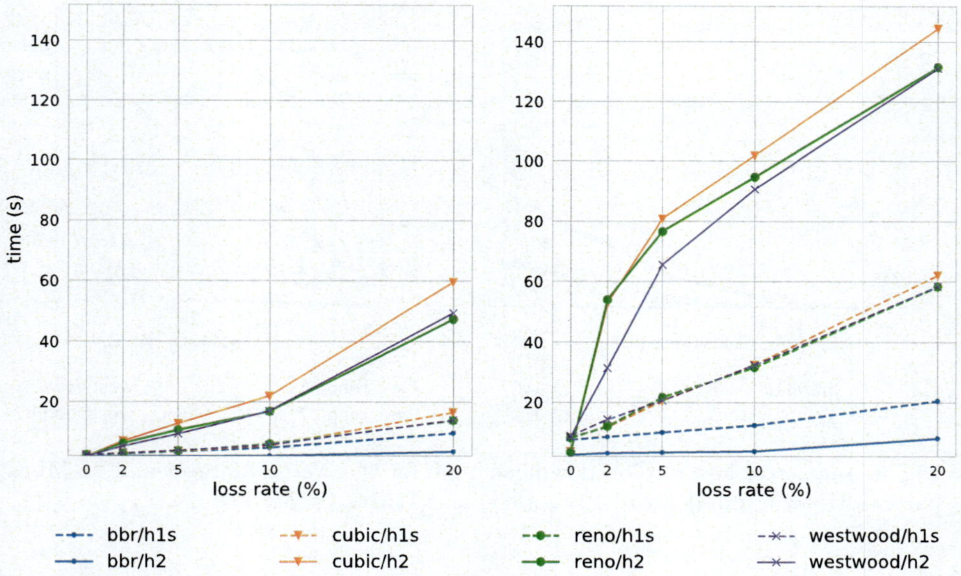

Fig. 7. Impact of loss rate on download time for the *Twitter* dataset and 20 Mbps of bandwidth for a link-delay of 10 ms (left) and 125 ms (right).

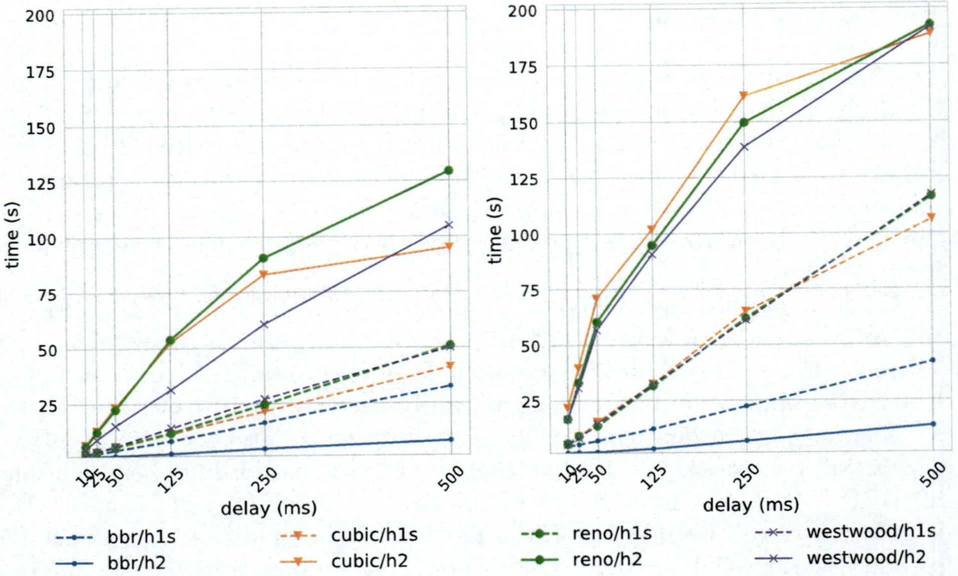

Fig. 8. Impact of delay for the *Twitter* dataset and 20 Mbps of bandwidth for a packet loss rate of 2% (left) and 10% (right).

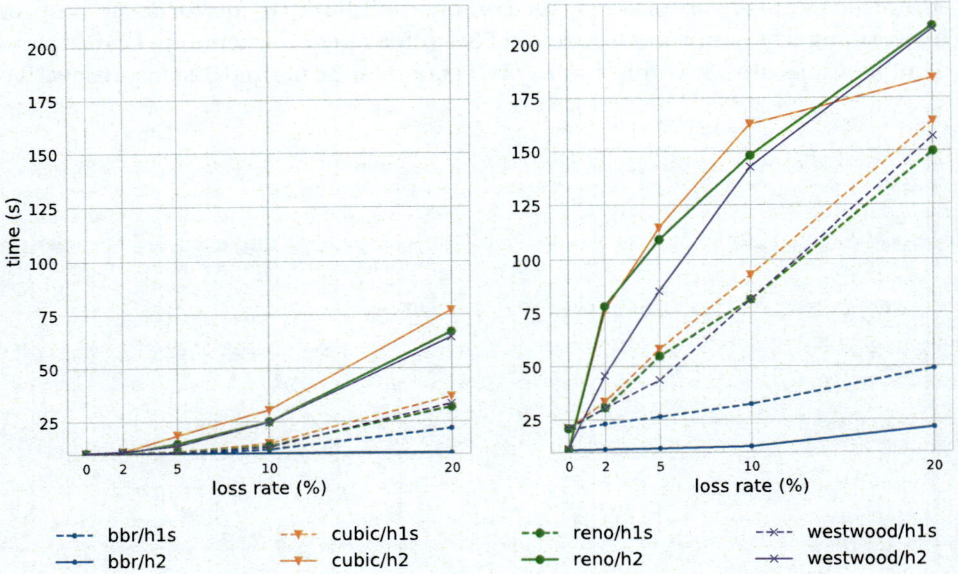

Fig. 9. Impact of loss rate on download time for the *Netflix* dataset and 20 Mbps of bandwidth for a link-delay of 10 ms (left) and 125 ms (right).

In the last case, it is interesting to note that the second-best is h1/Cubic (+15%), but it is still h1 the best performing application protocol.

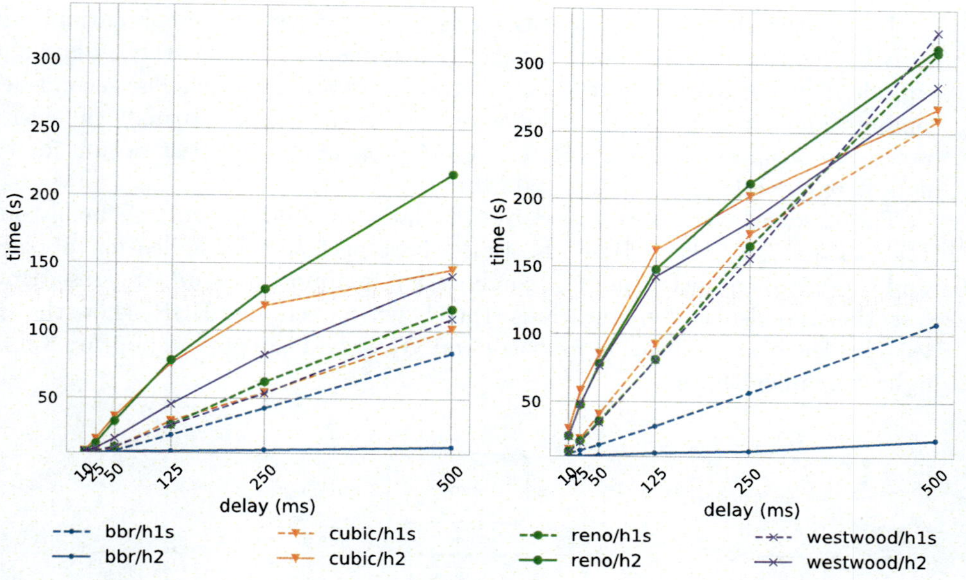

Fig. 10. Impact of delay for the *Netflix* dataset and 20 Mbps of bandwidth for a packet loss rate of 2% (left) and 10% (right).

Finally, recalling the scenario of 20 MB of bandwidth, h1s/BBR is the best combination when using a link delay of 10 ms (−15% on average) up to 10% of packet loss. Then, differences grow up 58% (in the extreme case of 20% of packet loss). The 125 ms scenario shows h2/BBR as the best option, outperforming h1/Westwood by 70% on average. When not using BBR, there are no significant differences under low RTT conditions, but h1/Westwood becomes the best strategy for a link of 125 ms of delay (−11.7%).

5.5 Summary by Dataset

In this section, we summarize the best performing combinations of an application protocol (i.e. HTTP version) and congestion control technique at the transport layer. We show them in the form of decision trees that consider the studied configurations. For each tree, the leaves include the best and second-best settings. We decided to show this information due to most of the best performing combinations use BBR as congestion control. However, BBR is not widely adopted, so the second configuration is based on congestion control techniques commonly used by Linux systems.

When considering the *continuous flow* dataset (Fig. 11) it is possible to observe that although BBR is the best performing congestion control algorithm, the general trend is that Westwood is almost always the second-best. Besides, it is possible to state that (in general) h2 is preferable over h1 over low loss rate links (and vice-versa for 10% of packet loss rate).

The picture for the *flags* dataset has a precise winning combination, that is, BBR/h2 (as we mentioned earlier, this dataset is by the way adapted to explode h2's features). Figure 12 shows the corresponding tree. The second-best combinations show h2 as the winner on low latency links (10 ms) while h1 is preferable on links of 125 ms. Here, CUBIC appears as the best option for the highest bandwidth link studied (20 Mbps).

The cases of the Twitter and Netflix datasets show interesting behaviours (Figs. 13 and 14). Using BBR, h2 is the best performing application protocol (and these two comprise the best combination in almost all studied cases). However, the second-best configurations (that is when excluding BBR) show that h1 becomes the best application protocol and Westwood the prefered option for the highest bandwidth links.

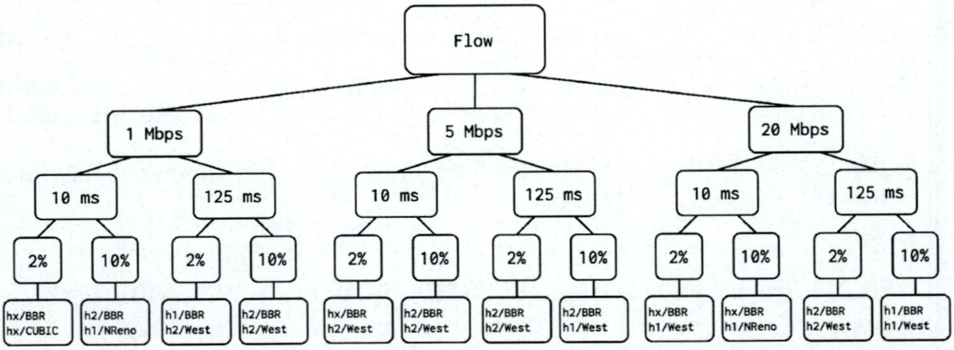

Fig. 11. Decision tree based on the results for the first and second best combination for the *continuous flow* dataset.

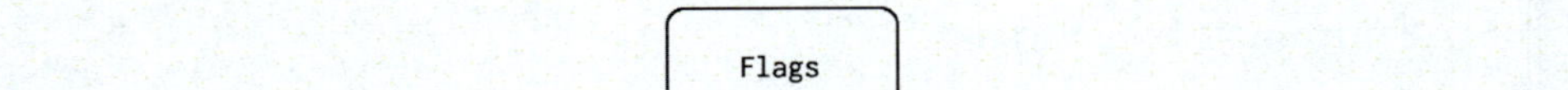

Fig. 12. Decision tree based on the results for the first and second best combination for the *flags* dataset.

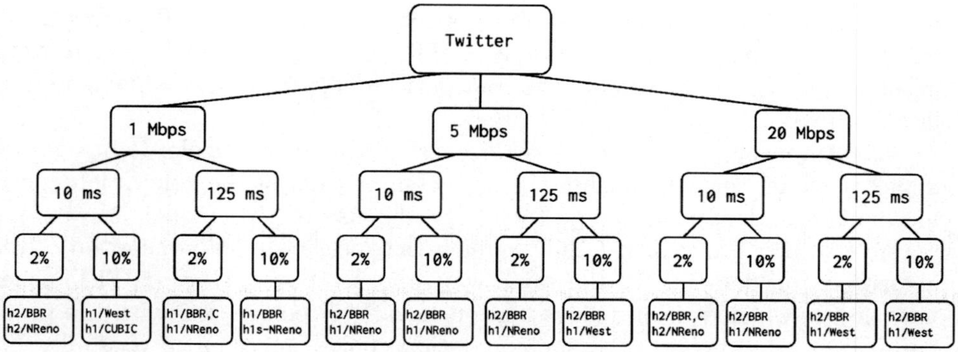

Fig. 13. Decision tree based on the results for the first and second best combination for *Twitter* dataset.

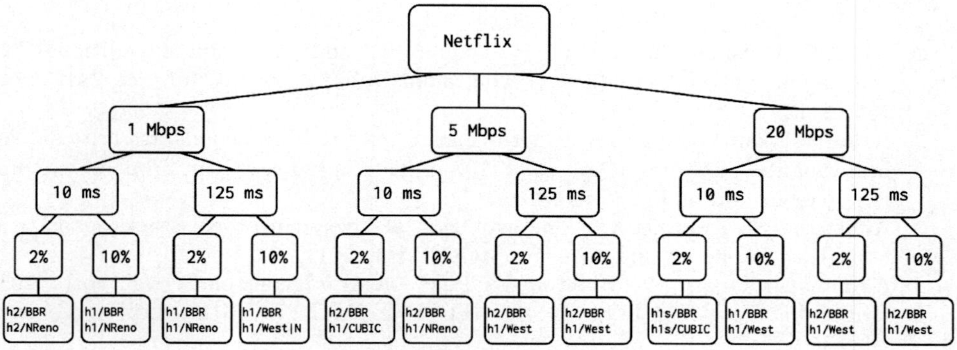

Fig. 14. Decision tree based on the results for the first and second best combination for *Netflix* dataset.

These results raise the following questions: is h2 worse that h1 as many studies claim, that triggered the development of HTTP/3 or it is just a matter of choosing the right congestion control algorithm? On the other hand, some studies show that a single BBR connection may monopolize an entire link, so it is necessary to control its application scenarios. All these results are a matter of future research that is out of the scope of this work.

6 Conclusions

The first observation of the overall experimentation is that there is an impact of the congestion control algorithm on the performance of HTTP/x when downloading different datasets. Consistently with previous work, BBR becomes the best congestion control mechanism in almost every setting, but the best HTTP version depends on the characteristics of the target site.

The most stable behaviour through the combinations happen when downloading the *continuous_flow* dataset. As we mentioned above, this is an expected result due to both application protocols requires only one TCP connection to handle this request, so the behaviour of HTTP/x is quite similar. As an overall observation, HTTP/2 becomes the best performing option over reliable and low latency links.

As future work, we propose to characterize a broader range of web pages and depict a model that establishes the best combination of the congestion control mechanism and application protocol. By the use of this model, it would be possible to propose a *hybrid* TCP version that uses different congestion control algorithms on demand. In the same way, and a more *intelligent* HTTP client may decide which version to use and opens a TCP connection with the proper parameters to indicate a specific congestion control mechanism to apply. This proposal may be extended to consider TLSv1.3 and HTTP/3 once it is deployed.

References

1. Abed, G.A., Ismail, M., Jumari, K.: Exploration and evaluation of traditional TCP congestion control techniques. J. King Saud Univ. - Comput. Inf. Sci. **24**(2), 145–155 (2012)
2. Akamai. Akamai online retail performance report: Milliseconds are critical. Web performance analytics show even 100-millisecond delays can impact customer engagement and online revenues (2017)
3. Akhshabi, S., Dovrolis, C.: The evolution of layered protocol stacks leads to an hourglass-shaped architecture. In: SIGCOMM (2011)
4. Allman, M., Paxson, V., Blanton, E.: TCP congestion control. Tech Reports (2009)
5. Benjamin, D.: Using TLS 1.3 with HTTP/2. RFC 8740, RFC Editor, February 2020
6. Bishop, M.: Hypertext transfer protocol version 3 (HTTP/3). Internet-Draft draft-ietf-quic-http-29, IETF Secretariat, June 2020. http://www.ietf.org/internet-drafts/draft-ietf-quic-http-29.txt
7. Brylinski, A., Bhattacharjya, A.: Overview of HTTP/2. In: ICC 2017: Proceedings of the Second International Conference on Internet of Things, Data and Cloud Computing (2017)
8. Callegari, C., Giordano, S., Pagano, M., Pepe, T.: A survey of congestion control mechanisms in Linux TCP. In: Vishnevsky, V., Kozyrev, D., Larionov, A. (eds.) Distributed Computer and Communication Networks, pp. 28–42. Springer International Publishing (2014)
9. Casetti, C., Gerla, M., Mascolo, S., Sanadidi, M.Y., Wang, R.: TCP Westwood: end-to-end congestion control for wired/wireless networks. Wirel. Netw. **8**, 467–479 (2002)
10. de Saxcé, H., Oprescu, I., Chen, Y.: Is HTTP/2 really faster than HTTP/1.1? In: IEEE Conference on Computer Communications Workshops (INFOCOM), pp. 293–299 (2015)
11. Elkhatib, Y., Tyson, G., Welzl, M.: Can SPDY really make the web faster? In: 2014 IFIP Networking Conference, pp. 1–9, June 2014
12. Goel, U., Steiner, M., Wittie, M.P., Ludin, S., Flack, M.: Domain-Sharding for faster HTTP/2 in lossy cellular networks. *CoRR*, abs/1707.05836 (2017)
13. Grieco, L.A., Mascolo, S.: Performance evaluation and comparison of Westwood+, New Reno, and Vegas TCP congestion control. SIGCOMM Comput. Commun. Rev. **34**(2), 25–38 (2004)

14. Grigorik, I.: High Performance Browser Networking. Shroff Publishers & Distributors, Bengaluru (2014)
15. Ha, S., Rhee, I., Xu, L.: CUBIC: a new TCP-friendly highspeed TCP variant. ACM SIGOPS Oper. Syst. Rev. **42**(5), 64–74 (2008)
16. Iyengar, J., Thomson, M.: QUIC: a UDP-based multiplexed and secure transport. Internet-Draft draft-ietf-quic-transport-29, IETF Secretariat, June 2020. http://www.ietf.org/internet-drafts/draft-ietf-quic-transport-29.txt
17. Jacobson, V.: Congestion avoidance and control. Comput. Commun. Rev. **18**(4), 314–329 (1988)
18. Jaeger, B., Scholz, D., Raumer, D., Geyer, F., Carle, G.: Reproducible measurements of TCP BBR congestion control. Comput. Commun. **144**, 31–43 (2019)
19. Langley, A., Riddoch, A., Wilk, A., Vicente, A., Krasic, C., Zhang, D., Yang, F., Kouranov, F., Swett, I., Iyengar, J., et al.: The QUIC transport protocol: design and internet-scale deployment. In: Proceedings of the Conference of the ACM Special Interest Group on Data Communication, SIGCOMM 2017, pp. 183–196 (2017)
20. Postel, J.: User datagram protocol. STD 6, RFC Editor, August 1980. http://www.rfc-editor.org/rfc/rfc768.txt
21. Postel, J.: Transmission control protocol. STD 7, RFC Editor, September 1981. http://www.rfc-editor.org/rfc/rfc793.txt
22. Qureshi, B., Othman, M., Hamid, N.A.W.: Progress in various TCP variants. In: 2nd International Conference on Computer, Control and Communication, pp. 1–6 (2009)
23. Schurman, E., Brutlag, J.: Performance related changes and their user impact. In: Velocity Web Performance and Operations Conference (2009)
24. Turkovic, B., Kuipers, F.A., Uhlig, S.: Interactions between congestion control algorithms. In: Network Traffic Measurement and Analysis Conference (TMA), pp. 161–168 (2019)
25. Varvello, M., Schomp, K., Naylor, D., Blackburn, J., Finamore, A., Papagiannaki, K.: Is the web HTTP/2 yet? In: Passive and Active Measurements Conference (PAM) (2016)
26. Wang, X., Xiao, S., Balasubramanian, A., Krishnamurthy, A., Wetherall, D.: How speedy is SPDY? In: NSDI, vol. 1, p. 1 (2014)
27. Yang, P., Shao, J., Luo, W., Xu, L., Deogun, J., Lu, Y.: TCP congestion avoidance algorithm identification. IEEE/ACM Trans. Netw. **22**(4), 1311–1324 (2014)
28. Yonav, E.: Amazon found every 100 ms of latency cost them 1% in sales (2009)
29. Ziakis, C., Vlachopoulou, M., Kyrkoudis, T., Karagkiozidou, M.: Important factors for improving google search rank. Future Internet **11**, 32 (2019)

Wireless Networking

Design and Preliminary Functionality Test of Road Network Traffic Monitoring System Based on Indoor SDWMN In-band Architecture

Phoo Phoo Thet Lyar Tun(✉) and Chaodit Aswakul

Wireless Network and Future Internet Research Unit, Department of Electrical Engineering, Faculty of Engineering, Chulalongkorn University, Bangkok, Thailand
6170389421@student.chula.ac.th, chaodit.a@chula.ac.th

Abstract. Software-defined wireless mesh network (SDWMN) combines the functionalities of wireless mesh network (WMN) and software defined networking (SDN) to achieve the goal of being effectively manageable of WMN. In this paper, the indoor SDWMN in-band testbed is proposed and implemented. The design and implementation of indoor SDWMN in-band testbed proposed in this paper is the preliminary testbed for the future real outdoor SDWMN in-band testbed for road traffic monitoring system. The testing results of indoor SDWMN in-band discussed in this paper shows that the system can function properly and becomes ready for future testing in the real outdoor environment.

Keywords: Software defined wireless mesh network · Road traffic monitoring application · In-band

1 Introduction

Bangkok, the capital city of Thailand, the actual location of future testbed area, is ranked as the most congested city in Asia and the second most congested city globally [1]. To solve this problem of road traffic congestion, it is necessary to set up road traffic monitoring systems based on the various types of network infrastructures. Setting up the underlying network infrastructure requires two basic kinds of connectivities, namely wired connectivity and wireless connectivity. Wired connectivity to set up the network infrastructure costs a lot as the network becomes large. Therefore, the consideration of wireless connectivity is the best choice for this paper for cost-effectiveness. Among the wireless connectivities, in this paper, IEEE 802.11 standard Wi-Fi [2] connection is mainly used together with other wide coverage technologies e.g. 3G, 4G connections, LoRaWAN [3], Zigbee [4] and WiMAX [5]. For the IEEE 802.11 wireless network setting, there are two modes of setting up the network, namely the infrastructure mode and the ad-hoc mode. In this work, the ad-hoc mode is chosen for the consideration of being easily extendable of the network.

B. Ghita and S. Shiaeles (Eds.): INC 2020, LNNS 180, pp. 181–192, 2021.
https://doi.org/10.1007/978-3-030-64758-2_13

The multi-hop ad-hoc network, the wireless mesh network (WMN) [6] is used for the cost-effectiveness of this paper. There are three traditional distributed routing protocols used in WMN, namely, proactive protocol, reactive protocol, and hybrid protocol [7]. All these traditional wireless ad-hoc routings are by-design distributed. Therefore, there is no central point to be controlled and managed. However, such distributed routing behavior is difficult to be modified.

Software Defined Networking (SDN) [8] is composed of the three planes, which are (i) data plane (ii) control plane and (iii) application plane. The forwarding elements are located on the forwarding plane. In SDN, the forwarding elements need to follow the instruction from the SDN controller for the behavior of packet forwarding, and the forwarding elements cannot decide for itself how to forward the packets. In traditional networking, the routing table inside the forwarding elements lets them decide how to forward the packets. In SDN, the SDN controller, which is the brain of the whole network, is located on the control plane. SDN controller uses OpenFlow protocol at the southbound interface as the communication protocol between the control and data planes. Thus, SDN with the help of OpenFlow protocol [9] has in the past been proposed to eliminate the limitation of traditional WMN, and the feature of network programmability provided by SDN lets the network administrator modify the routing behavior easily by using high-level programming languages. SDN based networks are easy to be controlled, altered, and managed because of the powerful features of separating the control plane and data plane of the network.

There are two ways to set up the control plane for the SDN, the out-of-band architecture uses the dedicated control plane and the in-band architecture uses the shared control plane [10]. Software-defined wireless mesh network (SDWMN) [11] combines the functionalities of WMN and SDN to achieve the goal of being effectively manageable of WMN and the cost-effective WMN has been implemented as a real outdoor testbed for the near real-time road traffic monitoring application using Apache-Kafka [12]. In the SDWMN in-band testbed proposed in [11], some mesh nodes has four hops far from the SDN controller. So, the control plane is unstable for this work. The design proposed in this paper concerns about the reliability of the control plane.

In the literature of wireless SDN [13–22], various approaches have been proposed and tested in the simulated, emulated and indoor environments. The researchers of [11] and [12] have tested in the outdoor environment for their road traffic monitoring application. The work of [11] and [12] uses the SDWMN in-band architecture. The controller placement of the work [11] is at the gateway of the network. In this paper, the indoor SDWMN in-band testbed is proposed and implemented. The design and implementation of indoor SDWMN in-band testbed proposed in this paper is the preliminary testbed for the future real outdoor SDWMN in-band testbed for road traffic monitoring system. The results of indoor SDWMN in-band discussed in this paper shows that the system is ready for testing real outdoor SDWMN in-band testbed.

The rest of the paper is summarized as follows. Section 2 presents the design and implementation of indoor SDWMN in-band Architecture. Section 3 gives the discussion based on the results received from the indoor SDWMN in-band testbed. Finally, the paper is concluded in Sect. 4.

2 Design of Indoor SDWMN In-band Architecture

The design of SDWMN in-band testbed is based on the idea and work of [11]. However, the design proposed in this paper has the modifications on [11] such as the routing patterns and rerouting patterns of the control plane and data plane and controller placement so that the network can achieve an efficient control plane performance by reducing the hops between the controller and nodes. The road traffic monitoring application applied in the proposed testbed is based on the idea and work of [12].

2.1 Design of Indoor SDWMN In-band Testbed

The indoor SDWMN in-band testbed is depicted in Fig. 1. The topology consists of six wireless mesh nodes pi1, pi2, pi3, pi4, pi5 and pi6, two gateways gw1 and gw2, one Ryu controller, one Kafka external broker and one switch. This switch is used to connect between the two gateways, the Ryu controller and the Kafka external broker. Six raspberry pi's 3 model B [23] with Quad-Core CPU and 1-GByte RAM are used as the testbed's wireless mesh nodes. Raspberry Pi is a credit-card sized computer that can give the cost-effectiveness and computational power within a compact form factor. Intel R NUC7i7BNH [24] is used as the two gateways. Dual-band EDUP EP-AC1605 Wi-Fi USB adapter [25] with two omnidirectional antennas are attached to the mesh nodes and the gateways. Since the road traffic monitoring is applied to the indoor SDWMN-indoor testbed, it can be said that there are two layers of abstraction. First is the network layer that is SDWMN in-band layer, and the second is the application layer that the road traffic monitoring application is running on. Therefore, for example, the raspberry pi 3 model B is used as the wireless mesh node in the network layer and is used as the application layer's Kafka producer. Intel R NUC7i7BNH is used as the wireless gateways in the network layer, and it is used as the Kafka broker, Kafka consumer and Kafka producer in the application layer. All the software and hardware selection, the frequency band selection for the network, the location selected for the final outdoor testbed, the reason to choose whether in-band or out-of-band, and the road traffic monitoring application appliance in the testbed are based on the work of [11] and [12].

In [11], gateway gw2 has to pass through 4 hops to reach the Ryu controller, which is in gw1. By considering the control plane reliability improvement, in this design, the per-hop distance between the nodes and the Ryu controller is reduced. As shown in Fig. 1, pi1, pi2 and pi4 are going to gw1, gw1, in turn, goes to the Ryu controller and pi3, pi5 and pi6 are going to gw2, gw2, in turn, goes to Ryu controller to establish the control plane. It means, in this design, we consider mainly for the controller placement and control plane reliability improvement. The data plane traffics are the images captured by the pi cameras. The data plane traffics use the same routing patterns as in [11] to communicate to the Kafka external broker. Therefore, in this design, the SDWMN in-band network is used to run the the road traffic monitoring application. The indoor SDWMN in-band testbed has been carried out in the author's room, which is located in the Ban Ratchathewi Apartment, Bangkok. Figure 2 and Fig. 3 show the indoor equipment settings.

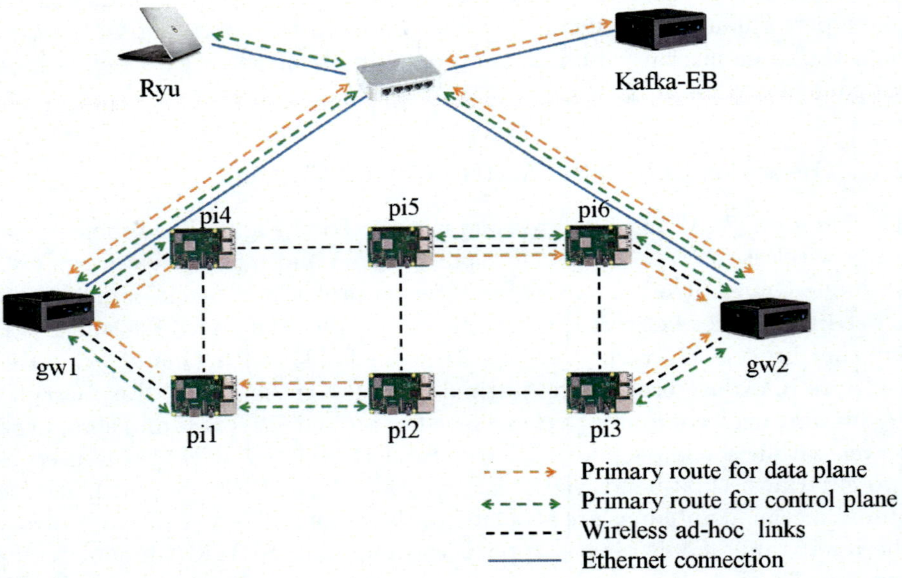

Fig. 1. Design of preliminary indoor SDWMN in-band testbed.

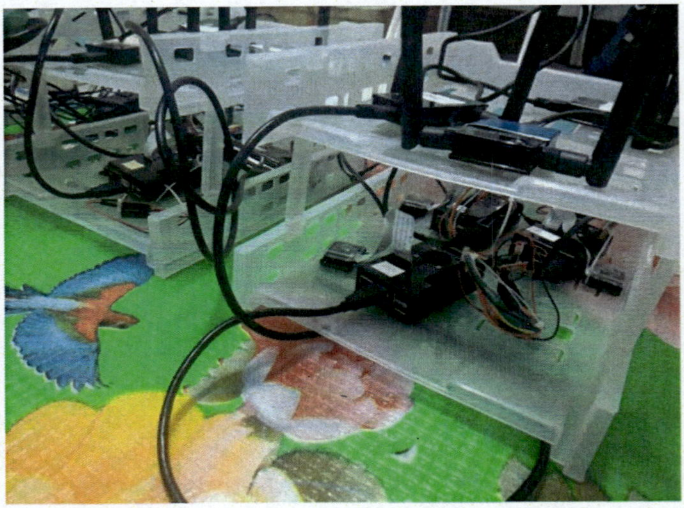

Fig. 2. Equipment setting of raspberry pi.

Fig. 3. Equipment setting of indoor SDWMN in-band testbed.

2.2 Implementation of Indoor SDWMN In-band Testbed

For the physical implementation of indoor SDWMN in-band testbed, the devices listed in Table 1 are used. The software needed for installation are listed in Table 2. The primary route patterns and the rerouting patterns for the data plane and control plane of the SDWMN in-band tested are shown in Fig. 4.

According to the design of the SDWMN in-band testbed presented in the previous section, the Ryu controller runs in the different devices. In contrast, in [11], the Ryu controller runs in the gateway. In this SDWMN in-band testbed, the two gateways, gw1 and gw2 in Fig. 1, serve as the relay functions for the control plane traffics and data plane traffics so that the mesh nodes can communicate to the Ryu controller and the Kafka external broker. Therefore, the gateway has two interfaces. The first one is the wireless ad-hoc interface to communicate with the wireless mesh nodes. The second one is the LAN interface to establish the LAN connection to the Ryu controller for the control plane traffic, and to the Kafka, external broker for the data plane traffics. All the necessary ideas, concepts, installation, and the implementation for the road traffic monitoring running on top of the SDWMN in-band testbed are the same as [12].

Table 1. Hardware list of indoor SDWMN in-band testbed.

Device	Functionality	Quantity
Intel ® NUC7i7BNH	As wireless gateway (Kafka producer/broker/consumer)	3
Raspberry pi 3 model B	As wireless mesh node (Kafka producer)	6
EDUP EP-AC1605 dual-band antenna	For the network reachability	8
Raspberry pi camera module	To capture images	6
Switch	To set up LAN connection between gateways, Ryu controller and Kafka external broker	1
Laptop	To run Ryu controller	1

Table 2. Software list of SDWMN in-band testbed.

Software	Function	Installed node
Open Vswitch	Virtual OpenFlow switch	Gateways and wireless mesh nodes
RYU [27]	SDN controller	Dedicated laptop
Ubuntu Mate 16.04 (32 bit) [28]	Linux operating system	Wireless mesh nodes (Kafka producers)
Ubuntu (32 bit) [29]	Linux operating system	Gateways, Laptop for Ryu controller, Kafka external broker

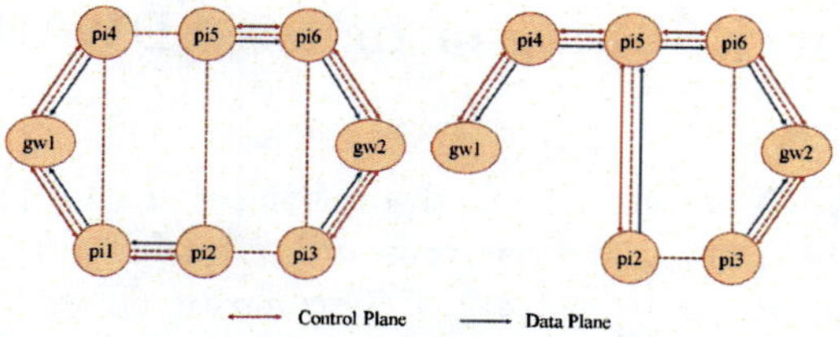

(a) pi1 unreachable to Ryu controller

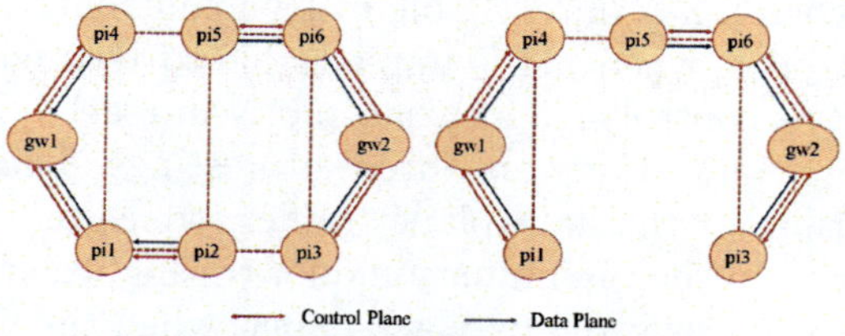

(b) pi2 unreachable to Ryu controller

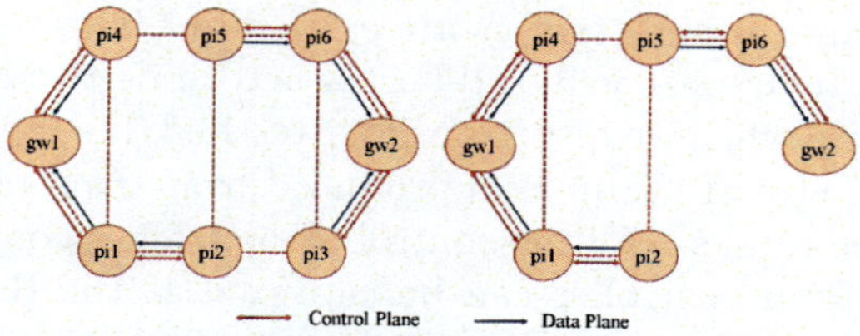

(c) pi3 unreachable to Ryu controller

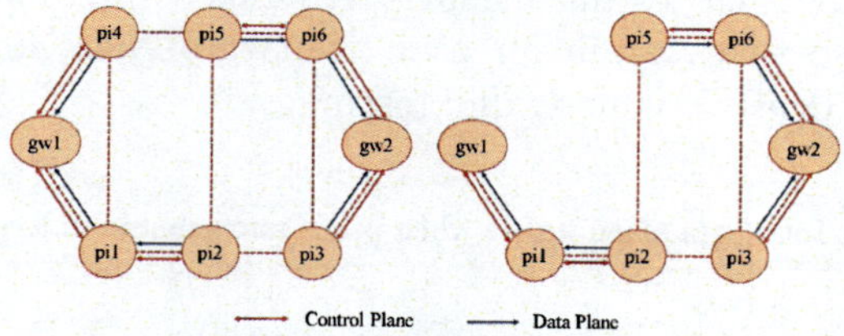

(d) pi4 unreachable to Ryu controller

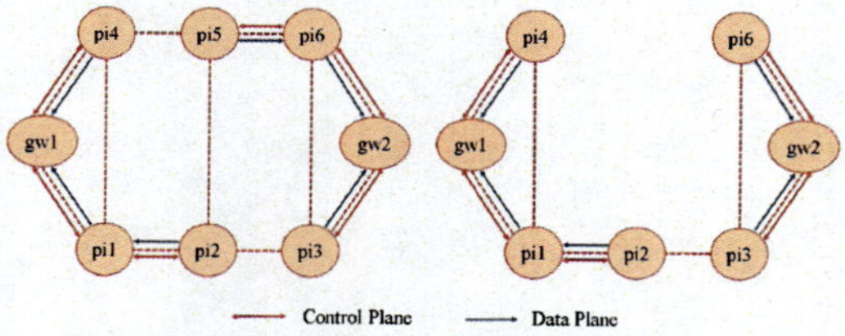

(e) pi5 unreachable to Ryu controller

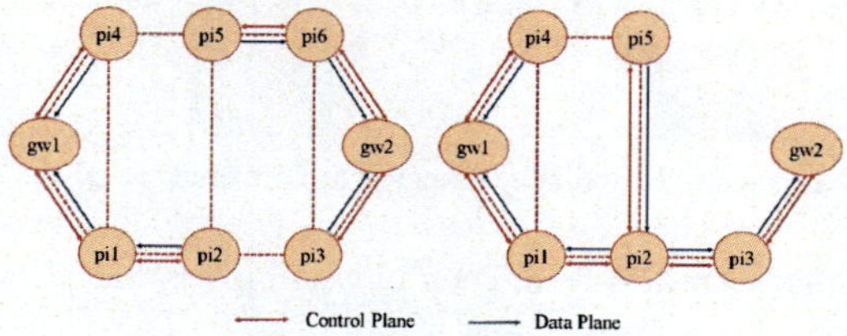

(f) pi6 unreachable to Ryu controller

Fig. 4. Routing patterns and rerouting patterns of indoor SDWMN in-band testbed.

3 Functionality Testing of Indoor SDWMN In-band Testbed

The indoor SDWMN in-band testbed applying road traffic monitoring application is tested to test the routing patterns and rerouting patterns of the data plane and control plane, there are a total of 7 scenarios. The testing scenarios range from when all pi's are reachable to Ryu controller, when pi1 is unreachable to Ryu controller, when pi2 is unreachable to Ryu controller, when pi3 is unreachable to Ryu controller, when pi4 is unreachable to Ryu controller, when pi5 is unreachable to Ryu controller and when pi6 is unreachable to Ryu controller. For each testing scenario, there are three sub-testings. These three sub-testings are aimed at testing for the data plane traffics, which are the images captured by the pi's cameras. In the preliminary indoor testing, the three sub-testings are when the pi's camera is covered by black paper, when the pi's camera is covered by white paper and when the pi's camera is uncovered.

The reason why the sub-testings are carried out is to test for the differences of the image size. Image size will be different according to the daytime, nighttime and traffic intensity when we test in outdoor testbed due to the compression used in [12]. The total size of the images produced from each pi can be captured at the gateways. The total size of the control plane traffics communicating between the pi's and the Ryu controller can be captured at the Ryu controller. In the preliminary indoor testing, the seven testing scenarios are tested separately. For each sub-testing of each testing scenario, it takes 3 min. The total control plane traffic testing also takes 3 min for each scenario. There are also seven testing scenarios for the control plane traffic testing.

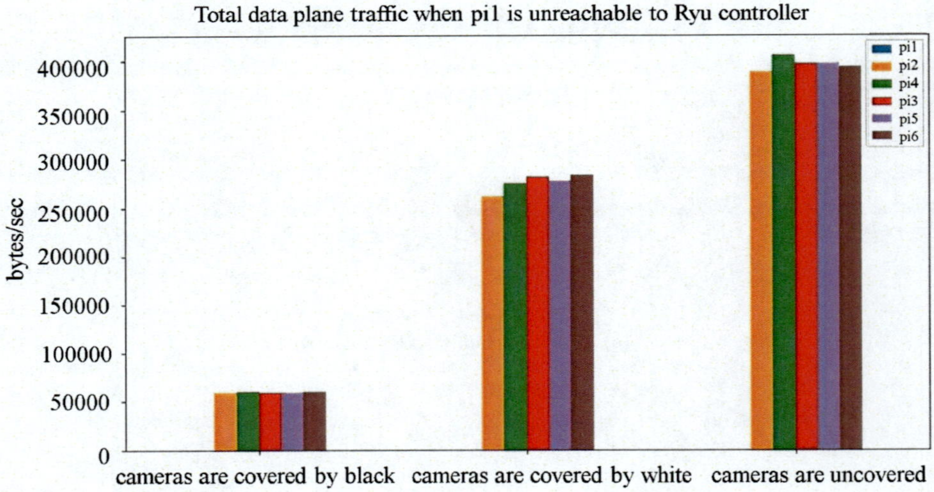

Fig. 5. Total data plane traffics from each pi when pi1 is unreachable to Ryu controller.

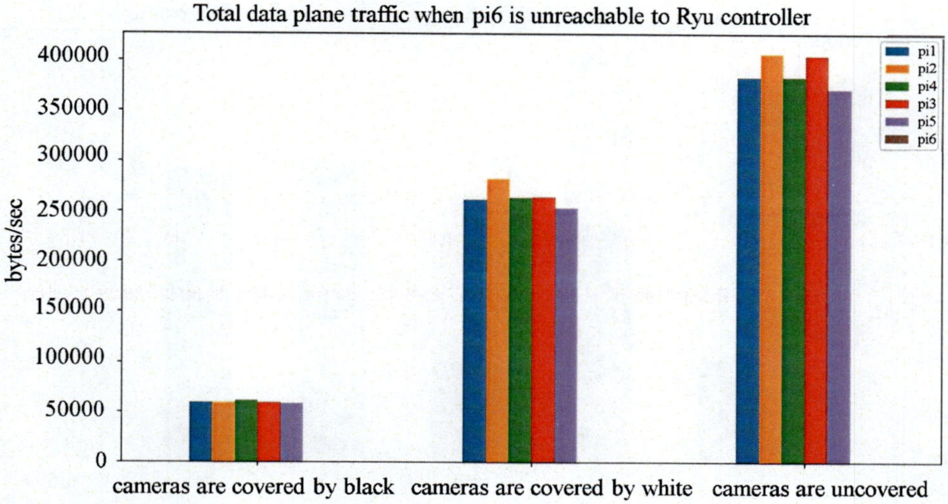

Fig. 6. Total data plane traffics from each pi when pi6 is unreachable to Ryu controller.

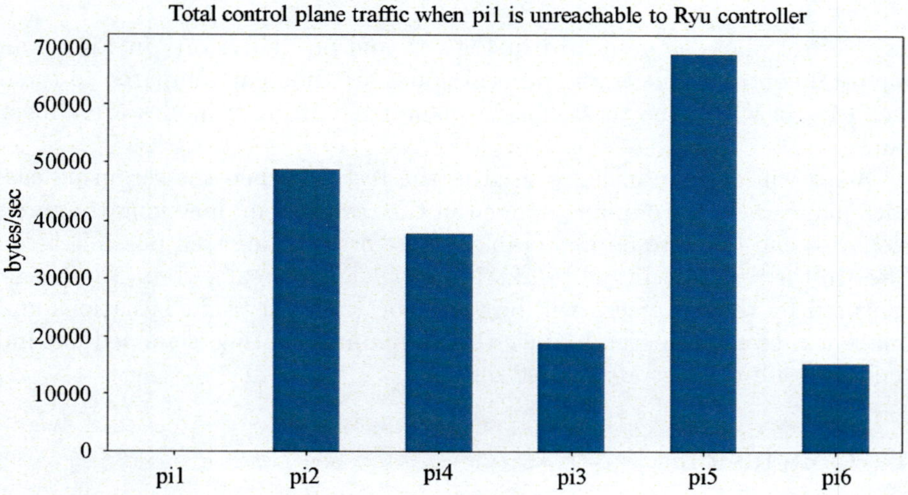

Fig. 7. Total control plane traffics from each pi when pi1 is unreachable to Ryu controller.

Even though pi1and pi6 are unreachable to Ryu controller, both the control plane traffic and data plane traffic of pi2 and pi5 are still reachable to the gateways as shown in Figs. 5, 6, 7 and 8. Since the rerouting takes a few seconds with the same setting as in [11], the total data plane traffic from pi2 and pi5 is lower than in the case of pi1 and pi6 unreachable cases as presented in Figs. 5 and 6. In the scenario of pi1 and pi6 unreachability, the Ryu controller has to assign the rerouting flow rules to pi2 and pi5 so that they can still establish

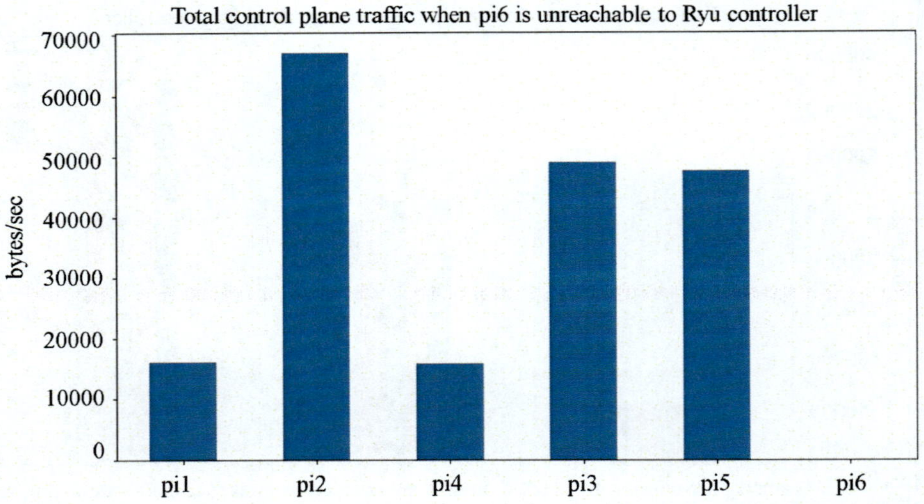

Fig. 8. Total control plane traffics from each pi when pi6 is unreachable to Ryu controller.

the control plane as described in Figs. 4a and 5d. Therefore, in Figs. 7 and 8, the total control plane traffic from pi nodes are different compared to the other scenarios in which the total control plane traffic from pi nodes are almost the same.

Since wifi ad-hoc links are used in the testing, when the rerouting cases of data plane and control plane happen in the real system, how long the rerouting scenarios can take are unpredictable. Therefore all the seven possible rerouting cases in the real system are tested. Moreover, in the real system, the image size captured by the pi camera will be varied on the road traffic situations and the image compression method. Since the preliminary testing is carried out indoor, the three sub-testings are carried out.

4 Conclusion

In this paper, the design and preliminary functionality testing of SDWMN in-band testbed applying road traffic monitoring application is proposed. The proposed design in this paper aims to achieve the better control plane reliability by reducing the hops between the mesh nodes and the Ryu controller. The indoor testbed is tested before testing the outdoor testbed which is the future work to check whether the routing patterns of the network, the rerouting scenarios after the failure of the primary routes, and the road traffic monitoring application applied in the testbed are able to work correctly or not. According to the results obtained from the SDWMN in-band testbed, the routing and rerouting application is confirmed to function properly. For the data plane testing, three sub-testings are tested. The testing results of indoor SDWMN in-band discussed in this paper shows that the system can function properly and becomes ready for future testing in the real outdoor environment.

Acknowledgment. This work was supported by the Asi@Connect's Data-Centric IoT-Cloud Service Platform for Smart Communities (IoTcloudServe@tEIN) project with grant contract ACA 2016/376-562 under the umbrella of Smart-Mobility@Chula demonstration site.

References

1. T. N. TomTom N. V. Dutch Company, Amsterdam: Tomtom traffic index: Bangkok, Thailand. https://www.tomtom.com/engb/trafficindex/city/bangkok. Accessed May 2020
2. Wi-Fi. https://www.wi-fi.org/. Accessed May 2020
3. LoRaWan. https://lora-alliance.org/. Accessed May 2020
4. ZigBee. https://www.zigbee.org/. Accessed May 2020
5. WiMAX. http://wimaxforum.org/. Accessed May 2020
6. Seyedzadegan, M., Othman, M., Ali, B.M., Subramaniam, S.K.: Wireless mesh networks: WMN overview, WMN architecture. In: International Conference on Communication Engineering and Networks IPCSIT, vol. 19, pp. 12–18 (2011)
7. Campista, M.E.M., Esposito, P.M., Moraes, I.M., Costa, L.H.M.K., Duarte, O.C.M.B., Passos, D.G., De Albuquerque, C.V.N., Saade, D.C.M., Rubinstein, M.G.: Routing metrics and protocols for wireless mesh networks. IEEE Netw. **22**(1), 6–12 (2008)
8. Open Network Foundation: Software-defined networking: the new norm for networks. Tech. rep., Open Networking Foundation (April 2012)
9. McKeown, N., Anderson, T., Balakrishnan, H., Parulkar, G., Peterson, L., Rexford, J., Shenker, S., Turner, J.: OpenFlow: enabling innovation in campus networks. Comput. Commun. Rev. **38**, 69–74 (2008)
10. Jalili, A., Nazari, H., Namvarasl, S., Keshtgari, M.: A comprehensive analysis on control plane deployment in SDN: in-band versus out-of-band solutions, pp. 1025–1031 (December 2017)
11. Htet, S.Y.: Design and implementation of medium-range outdoor wireless mesh-network with open-flow in raspberry pi. Master's thesis, Chulalongkorn University, Bangkok, Thailand (2018)
12. Htut, A.M.: Development of near real-time wireless image streaming cloud with Apache Kafka for road traffic monitoring application. Master's thesis, Chulalongkorn University, Bangkok, Thailand (2019)
13. Dely, P., Kassler, A., Bayer, N.: OpenFlow for wireless mesh networks. In: 2011 Proceedings of 20th International Conference on Computer Communications and Networks (ICCCN), pp. 1–6 (2011)
14. Detti, A., Pisa, C., Salsano, S., Blefari-Melazzi, N.: Wireless mesh software defined networks(WMSDN). In: 2013 IEEE 9th International Conference on Wireless and Mobile Computing, Networking and Communications (WiMob), pp. 89–95 (2013)
15. Lee, W.J., Shin, J.W., Lee, H.Y., Chung, M.Y.: Testbed implementation for routing WLAN traffic in software defined wireless mesh network. In: 2016 Eighth International Conference on Ubiquitous and Future Networks (ICUFN), pp. 1052–1055 (2016)

16. Yang, H., Chen, B., Fu, P.: OpenFlow-based load balancing for wireless mesh network, pp. 368–379 (January 2015)
17. Patil, P., Hakiri, A., Barve, Y., Gokhale, A.: Enabling software-defined networking for wireless mesh networks in smart environments. In: 2016 IEEE 15th International Symposium on Network Computing and Applications (NCA), pp. 153–157 (2016)
18. Arun, K.P., Chakraborty, A., Manoj, B.S.: Communication overhead of an openflow wireless mesh network. In: 2014 IEEE International Conference on Advanced Networks and Telecommunications Systems (ANTS), pp. 1–6 (2014)
19. Mamidi, A.V., Babu, S., Manoj, B.S.: Dynamic multi-hop switch handoffs in software defined wireless mesh networks. In: 2015 IEEE International Conference on Advanced Networks and Telecommunications Systems (ANTS), pp. 1–6 (2015)
20. Bao, K., Matyjas, J.D., Hu, F., Kumar, S.: Intelligent software-defined mesh networks with link-failure adaptive traffic balancing. IEEE Trans. Cogn. Commun. Netw. **4**(2), 266–276 (2018)
21. Sriramulu, R.K.: Constructing dynamic ad-hoc emergency networks using software-defined wireless mesh networks. Master's thesis, San Jose State University, San Jose, CA, USA (2018)
22. Yu, C., Yang, Z., Chen, X., Yang, J.: Scalable video transmission in software defined wireless mesh network. In: 2018 4th IEEE Conference on Network Softwarization and Workshops (NetSoft), pp. 456–461 (2018)
23. Raspberry Pi 3. https://www.raspberrypi.org/. Accessed May 2020
24. Intel®NUC 7i7BNH. https://www.intel.com/content/www/us/en/products/boards-kits/nuc/kits/nuc7i7bnh.html. Accessed Jan 2019
25. EDUP EP-AC1605. http://www.szedup.com/. Accessed May 2020
26. Open vSwitch. http://openvswitch.org. Accessed May 2020
27. RYU SDN Framework. https://osrg.github.io/ryu-book/en/Ryubook.pdf. Accessed May 2020
28. Ubuntu-Mate. https://ubuntu-mate.org/. Accessed May 2020
29. Ubuntu. https://www.ubuntu.com/. Accessed May 2020

Power Optimized Source-Based-Jamming for Secure Transmission through Untrusted AF Relays

P. M. Shemi[1,2](✉), M. G. Jibukumar[1], and M. A. Ali[3]

[1] Division of Electronics, School of Engineering, Cochin University of Science and Technology, Cochin, India
shemipm@gmail.com, jibukumar@cusat.ac.in

[2] Department of Electronics, MES College Marampally, Aluva, Kerala, India

[3] Department of Computer Applications, Government Engineering College, Trissur, India
ali@gectcr.ac.in

Abstract. In this paper, a power optimized source-based-jamming scheme is proposed to improve the secrecy of cooperative networks comprising of multiple untrusted amplify-and-forward relays in the presence of an external eavesdropper. Nelder-Mead gradient-free optimization algorithm is used for power allocation. The secrecy performance of untrusted relaying scheme is compared with worst case scenario; where the relays and external eavesdropper are assumed to be cooperating with each other. For performance comparison, optimization algorithms such as two-dimensional exhaustive search and gradient-based methods for symmetric and asymmetric relay positions have been derived. The complexity analysis of the proposed algorithm and its performance comparison with equal power allocation (EPA) strategy are also studied. Numerical results reveal that the proposed scheme outperforms other optimization methods, EPA and worst case jamming strategies.

Keywords: Nelder-Mead method · Optimal power allocation · Secrecy rate · Source-based-jamming

1 Introduction

Cooperative communication is a promising method used for improving the performance of physical-layer-security (PLS) against eavesdropping. The different PLS solutions that include cooperative relaying and cooperative jamming techniques have drawn great interests in recent research. Cooperative relaying increases the reliability of transmission and extends the coverage of wireless networks [1]. It also overcomes the problems with fading in wireless networks. Relay selection techniques can be used to select the best relay from multiple relaying candidates to cooperate with a communication link, which helps in reducing the computational complexity during the signal processing operation of the wireless networks. Cooperative jamming (CJ) is an alternate solution where relay

B. Ghita and S. Shiaeles (Eds.): INC 2020, LNNS 180, pp. 193–206, 2021.
https://doi.org/10.1007/978-3-030-64758-2_14

selection techniques fail to guarantee perfect secrecy. CJ is classified as destination-based-jamming (DBJ), source-based-jamming (SBJ), friendly jammer based jamming and hybrid jamming, depending on which node/nodes are transmitting the jamming signals. The secrecy rate, which quantifies the maximal rate of transmission at which the eavesdropper cannot decode any of the information, is taken as the main performance metric of PLS.

Relays can be of trusted or untrusted in cooperative networks. In trusted scenario, secure communication between source and destination occurs with the help of relays even in the presence of eavesdroppers. In practical scenarios like public adhoc networks or in heterogeneous networks, the relays used for connectivity may not be authenticated; they have a lower security clearance in the network; hence they are not considered trusted. In such cases, secrecy of the information transmitted via relay nodes need to be protected, despite the fact that the relay is a cooperating node. The relays can thus be observed as beneficial nodes as well as potential eavesdroppers, hence untrusted [2]. The untrusted relaying system was first investigated by He and Yener in [3], where DBJ with amplify-and-forward (AF) and compress-and-forward (CF) relaying schemes is performed to achieve positive secrecy.

The secrecy of the system that employs both cooperative relaying and cooperative jamming schemes is further improved by power optimization techniques. A survey of optimization approaches for wireless PLS in [4] discusses various topics on PLS designs, performance metrics, different categories of optimization problems etc. In [5], power optimization by a gradient-free optimization algorithm in trusted AF relaying networks that employs SBJ and Ant Colony Optimization based relay selection scheme is presented. The authors in [6] presented optimal power allocation (OPA) in the presence of a single untrusted relay whereas [7] presented the impact of OPA with multiple untrusted relays and potential eavesdroppers; where DBJ is used. Compared to other CJ methods, DBJ can be implemented easily as destination can remove the jamming signal from the prior information of the same [8]. The source - destination direct link cannot be considered in DBJ schemes that employ half-duplex relaying system. Only with SBJ, the direct link can be used and the flexibility of cooperation can be fully exploited. The SBJ techniques in [8] employ only untrusted relays for relaying and no external eavesdropper is assumed. The gradient-based (GB) optimization method used for secrecy rate maximization in [6, 8, 9], has problems with discontinuous functions, non-differentiable functions, mixed variable or large dimensionality functions. In [10], power optimization by gradient-free method in trusted AF cooperative relaying networks that employs hybrid jamming scheme is presented. But hybrid jamming is complicated in view of using two jamming signals at the source and relay nodes. Motivated by the above-mentioned observations, we proposed a gradient-free OPA strategy that employs an SBJ scheme to improve the security of untrusted AF relay networks, where source-destination link is assumed. The Nelder-Mead gradient-free algorithm overcomes the problems with GB power optimization method [11].

The main contributions of the work are outlined as:

A Nelder-Mead (N-M) based power allocation strategy that employs SBJ scheme is formulated for secrecy enhancement in two-hop untrusted AF cooperative networks in the presence of an external eavesdropper. The secrecy performance is evaluated under conventional and worst case untrusted scenarios. The complexity analysis and the performance comparison with other optimization algorithms and also with EPA strategy (CJ without power optimization) for symmetric and asymmetric relay positions are also studied.

2 System Description and Proposed Transmission Scheme

The system model shown in Fig. 1, consists of a source S, a destination D, N untrusted non-colluding randomly distributed AF relays represented by $\boldsymbol{R} = \{R_k \mid k = 1, 2, \ldots, N\}$, and a passive external eavesdropper E_e, who hide its existence in the network. The relays that operate in half-duplex mode, act as helpers of transmitting information as well as potential eavesdroppers, hence named as internal eavesdroppers. Each node employs omnidirectional antenna; the channels are modelled as Rayleigh flat fading. The source-destination link is considered in order to fully exploit the benefits of cooperation. Complete information transmission takes place in two time slots as time-division multiple-access is employed assuming total power of the system as P. The additive noise at any receiver is of the form of complex Gaussian random variable with zero mean and variance σ^2 and it is represented as CN (0, σ^2).

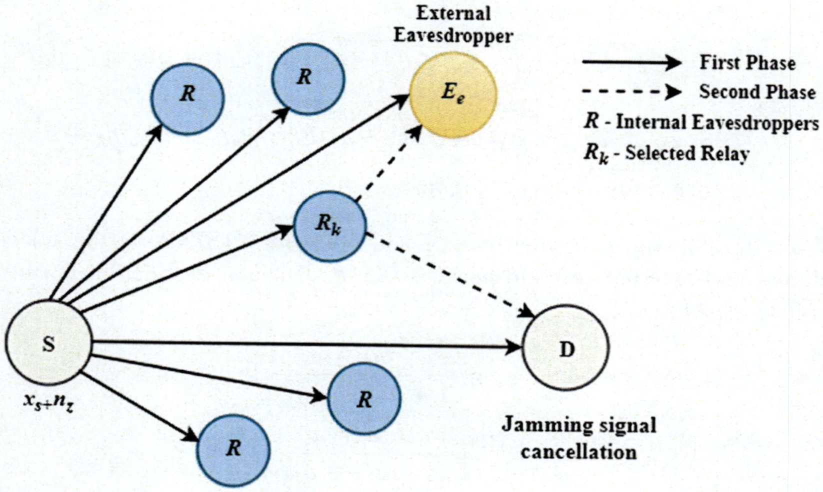

Fig. 1. System Model

An SBJ scheme, where the system allocates some of its source power to transmit the jamming signal to confound the internal/external eavesdroppers is proposed. The signal transmission involves broadcast and relaying phases. Denoting x_s and n_z as the information and jamming signals respectively, the signal transmitted from the source x is

$$x = \sqrt{aa_s P} x_s + \sqrt{a(1 - a_s)P} n_z \quad (1)$$

During the first transmission phase, the signal received at the internal and external eavesdroppers i, $i \in \{1, 2, \ldots, N, E_e\}$, and at the destination can be expressed as,

$$y_1^{(1)} = \sqrt{aa_s P} h_{si} x_s + \sqrt{a(1-a_s)P} h_{si} n_z + n_{i_1} \tag{2}$$

$$y_d^{(1)} = \sqrt{aa_s P} h_{sd} x_s + \sqrt{a(1-a_s)P} h_{sd} n_z + n_{d_1} = \sqrt{aa_s P} h_{sd} x_s + n_{d_1} \tag{3}$$

n_i is the additive noise at the node i, where $i \in \{R, E_e\}$

A relay k is selected prior to the second transmission phase. k, $k \in \{1,2,\ldots,N\}$ indicates the index of the selected relay R_k. For simplicity, the index of the relay is used in the equations. The selected relay k then amplifies the signal by an amplification factor g and broadcasts the message $x_k = g\, y_k$ to destination which is also received by the other untrusted relays and external eavesdropper E_e. g is expressed as

$$g = \sqrt{\frac{(1-a)P}{|h_{sk}|^2 aP + \sigma_k^2}} \tag{4}$$

The signal at the eavesdroppers i, $(i \neq k)$ and at the destination during the second transmission phase are

$$\begin{aligned} y_i^{(2)} &= g h_{ki} y_k + n_{i_2} \\ &= g\sqrt{aa_s P} h_{sk} h_{ki} x_s + g\sqrt{a(1-a_s)P} h_{sk} h_{ki} n_z + g h_{ki} n_k + n_{i_2} \end{aligned} \tag{5}$$

$$\begin{aligned} y_d^{(2)} &= g\sqrt{aa_s P} h_{sk} h_{kd} x_s + g\sqrt{a(1-a_s)P} h_{sk} h_{kd} n_z + g h_{kd} n_k + n_{d_2} \\ &= g\sqrt{aa_s P} h_{sk} h_{kd} x_s + g h_{kd} n_k + n_{d_2} \end{aligned} \tag{6}$$

From (2), the signal to interference and noise ratio (SINR) at the selected relay k, internal and external eavesdroppers i, $i \in \{R, E_e\}$ and destination during the first transmission phase is

$$\gamma_k = \frac{aa_s \gamma_{sk}}{1 + a(1-a_s)\gamma_{sk}} \tag{7}$$

$$\gamma_i^{(1)} = \frac{aa_s \gamma_{si}}{1 + a(1-a_s)\gamma_{si}} \tag{8}$$

$$\gamma_d^{(1)} = aa_s \gamma_{sd} \tag{9}$$

γ_{si}, γ_{sk} and γ_{sd} are the instantaneous SNR in the source-malicious node, source-selected relay and source-destination channels respectively.

The instantaneous SNR between two nodes i and j is mathematically expressed as

$$\gamma_{ij} = P\frac{|h_{ij}|^2}{\sigma_j^2} \tag{10}$$

where $i \in \{S, R\}$ and $j \in \{R, D, E_e\}$.

From (5) and (6), the SINR at the internal and external eavesdroppers i, $(i \neq k)$; and at the destination during the second transmission phase are

$$\gamma_i^{(2)} = \frac{a(1-a)a_s\gamma_{sk}\gamma_{ki}}{1 + a(1-a)(1-a_s)\gamma_{sk}\gamma_{ki} + a\gamma_{sk} + (1-a)\gamma_{ki}} \tag{11}$$

$$\gamma_d^{(2)} = \frac{a(1-a)a_s\gamma_{sk}\gamma_{kd}}{1 + a\gamma_{sk} + (1-a)\gamma_{kd}} \tag{12}$$

The overall SNR at the destination applying maximal ratio combing (MRC) is

$$\begin{aligned} \gamma_d &= \gamma_d^{(1)} + \gamma_d^{(2)} \\ &= aa_s\gamma_{sd} + \frac{a(1-a)a_s\gamma_{sk}\gamma_{kd}}{1 + a\gamma_{sk} + (1-a)\gamma_{kd}} \end{aligned} \tag{13}$$

All the noise variances are set equal for simplicity.

3 Secrecy Rate Analysis

The secrecy rate is used as the performance index here. In the proposed scenario, both untrusted relays (internal eavesdroppers) and external eavesdropper exist in the network at the same time. The untrusted relays act as both essential relays and malicious eavesdroppers, which can eavesdrop the information. In cooperative communication system, the instantaneous secrecy rate is computed by [7]

$$R_s = (R_D - R_E)^+ = \left[\frac{1}{2}\log_2(1+\gamma_d) - \frac{1}{2}\log_2(1+\gamma_E)\right]^+ \tag{14}$$

where $(x)^+ = \max\{0, x\}$;

γ_E is the amount of information leakage to malicious nodes; i.e., both untrusted relays and external eavesdropper. Since the power optimization method used here can allocate powers optimally between the source and relay nodes as well as between the information and jamming signals, $R_s \geq 0$ is achievable. Hence instantaneous secrecy rate in (14) is changed to [9]

$$R_s = \frac{1}{2}\log_2(1+\gamma_d) - \frac{1}{2}\log_2(1+\gamma_E) \tag{15}$$

3.1 Relay Selection

The optimal relay k is the one that gives maximum secrecy rate and it needs to satisfy

$$k^* = \arg\max_{1\leq k\leq N} R_s^{(k)} \tag{16}$$

The instantaneous CSI of each link has to be known for finding the relay of (16). Since in practical systems, it is difficult to use the optimal relay as we cannot get the information of external eavesdropper accurately, we go for suboptimal selection. A relay selection which avoids using the CSI of external eavesdropper is therefore used and the model without external eavesdropper is considered. If k is the selected relay, R_D and R_{Rk} are the rates at destination and selected relay respectively, the achievable secrecy rate $\left(R_{s_{wo\_Ee}}^{(k)}\right)$ is given as [12].

$$\begin{aligned} R_{s_{wo\_E_e}}^{(k)} &= \left[R_D - \underset{1 \le k \le N}{\arg\max} \left(R_{R_k} \right) \right]^+ \\ &= \left[\frac{1}{2}\log_2(1+\gamma_d) - \underset{1 \le k \le N}{\arg\max} \left(\frac{1}{2}\log_2(1+\gamma_k) \right) \right]^+ \\ &= \left[\frac{1}{2}\log_2\left(1 + aa_s\gamma_{sd} + \frac{a(1-a)a_s\gamma_{sk}\gamma_{kd}}{1+a\gamma_{sk}+(1-a)\gamma_{kd}}\right) - \underset{1 \le k \le N}{\arg\max} \left(\frac{1}{2}\log_2\left(1+\frac{aa_s\gamma_{sk}}{1+a(1-a_s)\gamma_{sk}}\right)\right) \right]^+ \end{aligned} \tag{17}$$

The relay selection criterion then becomes

$$k^* = \underset{1 \le k \le N}{\arg\max} \left(aa_s\gamma_{sd} + \frac{a(1-a)a_s\gamma_{sk}\gamma_{kd}}{1 + a\gamma_{sk} + (1-a)\gamma_{kd}} \right) \tag{18}$$

3.2 System Secrecy Rate

The secrecy rate for the proposed untrusted relaying network is obtained by (15). N untrusted relays and an external eavesdropper comprise a total of $(N + 1)$ malicious nodes. Here, the signals received by the malicious nodes during the first and second phases are considered separately. The amount of information leakage γ_E is the maximum of leakage to untrusted relays and external eavesdropper [7]

$$\gamma_E = \max_{1 \le i \le N+1, i \ne k} \left\{ \gamma_k, \gamma_i^{(1)}, \gamma_i^{(2)} \right\} \tag{19}$$

The secrecy rate with the selected relay is expressed as

$$\begin{aligned} R_s^{(k)}(a, a_s) = \Bigg\{ &\frac{1}{2}\log_2\left(1 + aa_s\gamma_{sd} + \frac{a(1-a)a_s\gamma_{sk}\gamma_{kd}}{1 + a\gamma_{sk} + (1-a)\gamma_{kd}}\right) \\ &- \frac{1}{2}\log_2\left(1 + \max_{1 \le i \le N+1, i \ne k} \left\{ \gamma_k, \gamma_i^{(1)}, \gamma_i^{(2)} \right\}\right) \Bigg\}^+ \end{aligned} \tag{20}$$

3.3 Secrecy Rate of Worst Case Scenario (WC)

The performance of the untrusted relaying scheme is compared with the worst case scenario, where the external eavesdropper and untrusted relays cooperate with each other. Here, the information leakage to the untrusted relays and the external eavesdropper is considered separately. The information received at the external eavesdropper in both phases can be combined. The amount of information leakage for the worst case scenario γ_{E_WC} is given by

$$\gamma_{E\_WC} = \max\left\{ \max_{1 \le i \le N, i \ne k}\left\{\gamma_k, \gamma_i^{(1)}, \gamma_i^{(2)}\right\}, \left(\gamma_{E_e}^{(1)} + \gamma_{E_e}^{(2)}\right)\right\} \tag{21}$$

where $\max_{1 \le i \le N, i \ne k}\left\{\gamma_k, \gamma_i^{(1)}, \gamma_i^{(2)}\right\}$ is the information leakage to the untrusted nodes R and $\left(\gamma_{E_e}^{(1)} + \gamma_{E_e}^{(2)}\right)$ is the leakage to the external eavesdropper E_e. The instantaneous secrecy rate is therefore given by,

$$R_{SWC}^{(k)}(a, a_s) = \left\{\frac{1}{2}\log_2\left(1 + aa_s\gamma_{sd} + \frac{a(1-a)a_s\gamma_{sk}\gamma_{kd}}{1 + a\gamma_{sk} + (1-a)\gamma_{kd}}\right) - \frac{1}{2}\log_2\left(1 + \max\left\{\max_{1 \le i \le N, i \ne k}\left\{\gamma_k, \gamma_i^{(1)}, \gamma_i^{(2)}\right\}, \left(\gamma_{E_e}^{(1)} + \gamma_{E_e}^{(2)}\right)\right\}\right)\right\}^+ \tag{22}$$

N-M method is applied to the functions in (20) and (22) to estimate the power optimal values for the proposed untrusted case and worst case scenario respectively. Since the N-M method finds the minimum of a function, the secrecy function is inverted before optimization.

4 Nelder-Mead Power Optimization Method (N-M)

J.A. Nelder and R. Mead formulated the N-M algorithm; a gradient-free optimization scheme widely used for unconstrained optimization of multidimensional non-linear functions [11]. It solves problems that gradient methods are difficult to solve. The algorithm minimizes a function of n variables depending on the function values at $(n + 1)$ vertices of a general simplex. Since the function to be optimized in this work depends on two variables a and a_s, the simplex has three vertices and therefore is a triangle. N-M algorithm starts with this simplex with vertices arranged based on the function values. After each iteration, the worst vertex is identified and replaced with a new vertex which forms a new simplex. The search continues till the simplex with minimum points is found [13].

The main steps involved in N-M algorithm are presented here for the completeness of the paper [14]. Let the function to be minimized be f which is our R_s. Since R_s is dependent on two power optimization parameters - a and a_s; algorithm for two-variable optimization is explained here. N-M algorithm for three-variable optimization for hybrid jamming scheme is presented in [10]. Four scalar factors namely reflection (δ), expansion (γ), contraction (ψ), and shrinkage (ρ) are defined for the N-M method. These parameters satisfy the following conditions.

$\delta > 0$; $\gamma > 1$, $\gamma > \delta$; $0 < \psi < 1$; $0 < \rho < 1$ [11].

The standard values used in most implementations are $\delta = 1$, $\psi = 0.5$, $\gamma = 2$, $\rho = 0.5$.

i. **Generate the simplex**: We prefer equal length simplex; assuming the length of the sides of the simplex is z $(z = 1)$. Let the initial guess (x_o, y_o) be the third vertex. The other two vertices (x, y) and (y, x) are found by adding a vector to the initial guess, where x and y are

$$y = \frac{z}{2\sqrt{2}}\left(\sqrt{3} - 1\right) \tag{23}$$

$$x = y + \frac{z}{\sqrt{2}} \tag{24}$$

The three points (x, y), (y, x), (x_o, y_o) correspond to the three vertices (x_1, y_1), (x_2, y_2), (x_3, y_3) of the simplex. After generating the initial simplex, evaluate the functions f_1, f_2 and f_3 at the corresponding vertices, where $f_i = f(x_i, y_i)$, $i = 1, 2, 3$. The vertices are then ordered such that $f_1 < f_2 < f_3$, so as to rank them as best (x_b), good (x_g) and worst (x_w) respectively as they denote the smallest, second largest and the largest function values. The initial guess is assumed as EPA values [0.5, 0.5].

ii. **Reflection**: The reflected point x_r is computed as

$$x_r = x_m + \delta(x_m - x_w) \tag{25}$$

where x_m is the middle point of line joining x_b and x_g given as,

$$x_m = \frac{x_b + x_g}{2} = \left(\frac{x_1 + x_2}{2}, \frac{y_1 + y_2}{2}\right) \tag{26}$$

The function value f_r at x_r is evaluated with $\delta = 1$.

iii. **Expansion**: If f_r is better than the best point, the reflection is successful and an expansion in the same direction is performed. The expanded point is

$$x_e = x_r + \gamma(x_r - x_m) \tag{27}$$

where the parameter γ is taken as 2. The function value f_e at x_e is evaluated; the iteration terminates after retaining x_e (if $f_e < f_r$) or x_r (if $f_e > f_r$).

iv. **Contraction**: If the reflected point is poorer than the worst point, a better point occurs between x_w and x_m and performs inside contraction,

$$x_c = x_m - \psi(x_m - x_w) \tag{28}$$

where the contraction parameter ψ is set to 0.5. If the function f_c at x_c is better than the worst point, keep the new point, else go to shrink (Step 5).

Outside contraction is done if the reflected point is not worse than the worst point, but worse than the good point x_g

$$x_o = x_m + \psi(x_m - x_w) \tag{29}$$

If the function f_o at x_o is better than the reflected point, keep the new point, else go to shrink (Step 5).

v. **Shrinking**: The best point is retained and shrinks the simplex. i.e., for all points except the best one, a new point is computed as

$$x_i = x_b + \rho(x_i - x_b) \tag{30}$$

where $i = 2, 3$ and the shrinkage parameter ρ is usually set as 0.5.

With these steps the iteration completes and a new simplex is formed. Then the process repeats with the new simplex.

5 Results and Discussion

To validate the effectiveness of the proposed secure transmission with OPA scheme, a two-dimensional coordinate system is considered with source and destination at (0, 0) and (10, 0) respectively and relay nodes randomly positioned between them. For the proposed scheme, the secrecy is analyzed by N-M algorithm and their performance is compared with EPA results and with other optimization methods like GB and ES algorithms. The simulation parameters considered for the analysis are as follows: the number of relay nodes $N = 20$, path loss $L = 3$, total transmit power $P = 30$ dBm, and power allocation factors $a = a_s = 0.5$ for EPA. The relay node is selected using (18).

Figure 2 gives the secrecy rate comparison among OPA/EPA schemes. Case 1 and 2 characterize the symmetric and asymmetric cases respectively depending on the relay position. Symmetric case represents the best relay position, where mostly relay at the center of the network model is selected. The other relay positions are asymmetric cases. Here, the relay near to source is considered for asymmetric case. It is understood from the figure that the secrecy increases with SNR and with jamming signals, since the overall SNR at the eavesdropper reduces. OPA achieves better secrecy at all SNRs as power is allocated based on the location of internal and external eavesdroppers. This effect is more predominant in asymmetric case. For both UT and WC symmetric case scenarios, EPA shows almost same performance as OPA; hence the variation among EPA and OPA is less for the entire SNR range. This is because of the fact that, when the relay at the center of the model is selected, source and relay nodes take almost equal power for transmission and that is similar to allocating 0.5 to a for EPA. For the case when the relay near to source is selected, the system requires less source power than relay power. Therefore, the variation among OPA and EPA is more for asymmetric case than that for the symmetric case and it is clearly understood from the figure. The secrecy rate changes at a rate of only 0.02 bits/s/Hz for Case 1, while it changes at a rate of 0.16 bits/s/Hz for Case 2 for every 2 dB change in SNR.

Figure 3 shows the variation of power allocation factors corresponding to the OPA results of Fig. 2 for the proposed untrusted scheme. The power allocation factors a and a_s depend on the position of internal and external eavesdroppers. For symmetric case, source and relay nodes take same powers for transmission. Hence a remains almost constant at 0.5; and the variation of a is between 0.52 and 0.56 for the entire SNR range. For the asymmetric case, source requires less power compared to relay ($a = 0.19$) since the relay near to source is considered. The external eavesdropper near to source is assumed for the analysis. Hence, more power should be given to jamming signals in

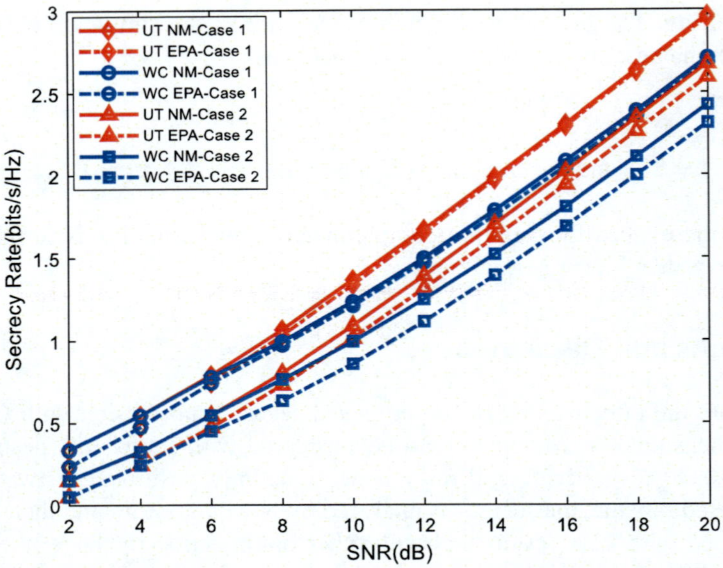

Fig. 2. OPA/EPA secrecy rate comparisons for untrusted and worst case scenarios

order to confuse the eavesdroppers during the first phase than the second phase. This is clear from the curves of a_s. Jamming powers $(1-a_s)$ of 0.5091 and 0.5284 are allocated for case 1 and case 2 respectively at 10 dB. Since secrecy is dependent on SNR, the variation is effective for SNRs of 10 dB and above.

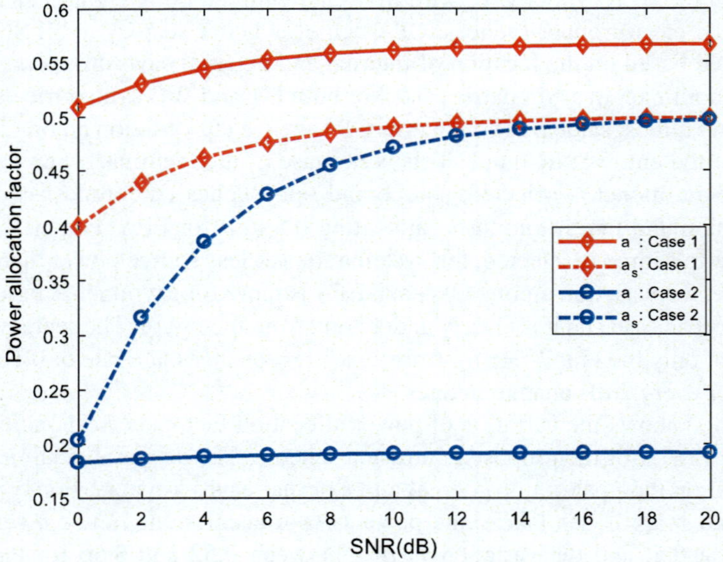

Fig. 3. Power allocation factors versus SNR for proposed untrusted case

Table 1 presents the performance comparison of the N-M method with GB and two-dimensional ES algorithms and EPA for symmetric and asymmetric relay positions. From the table, it is clear that the secrecy is highest with N-M method. The problem with gradient method is that it is time consuming to compute the function derivatives for a nonlinear function like secrecy rate. ES method, being the simplest of all methods produces accurate results, and the accuracy increases with the number of iterations m. But ES method is computationally inefficient and it is replaced with heuristic approach. The results of ES algorithm are obtained by matlab simulation and the surface plots for symmetric/asymmetric relay cases with $m = 20$ are given in Figs. 4(a) and (b) respectively. For the symmetric case, EPA and OPA secrecy rate values are almost same whereas for the asymmetric case, variation among them is higher. i.e., with N-M method, the variation is 0.074562 bits/s/Hz and 0.025478 bits/s/Hz for asymmetric and symmetric cases respectively.

Table 1. Performance comparison of power optimization methods and EPA strategy

Optimization Method	Symmetric case			Asymmetric case		
	a	a_s	R_s (bits/s/Hz)	a	a_s	R_s(bits/s/Hz)
N-M method	0.560966	0.490926	1.367478	0.191473	0.471595	1.090262
Gradient method	0.5604	0.4909	1.3654	0.1906	0.4716	1.0896
Exhaustive search	0.55	0.5	1.346	0.2	0.45	1.089
EPA	0.5	0.5	1.3380	0.5	0.5	1.0157

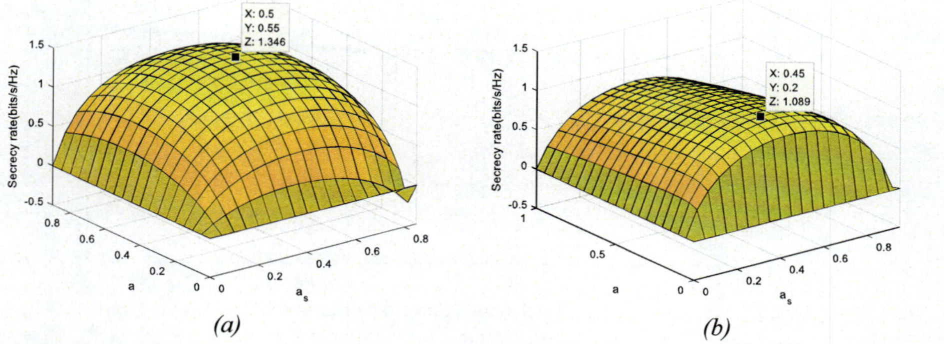

Fig. 4. Surface plot of secrecy rate versus power allocation factors a & a_s *(a)* Symmetric case *(b)* Asymmetric case

Figure 5 and Fig. 6 illustrate the secrecy rate and their power allocation factors between both untrusted cases in terms of source-relay distance. The relays are deployed between source and destination which are 10 m apart.

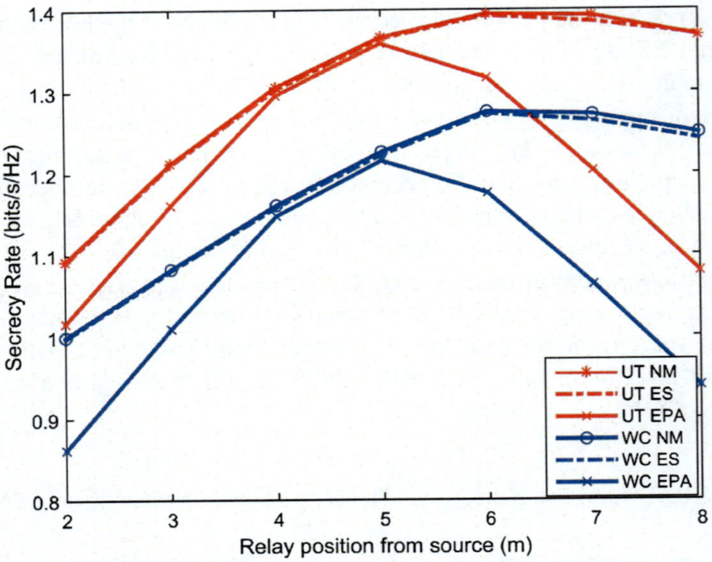

Fig. 5. OPA/EPA performance results among the untrusted and worst case scenarios in terms of relay distance from source

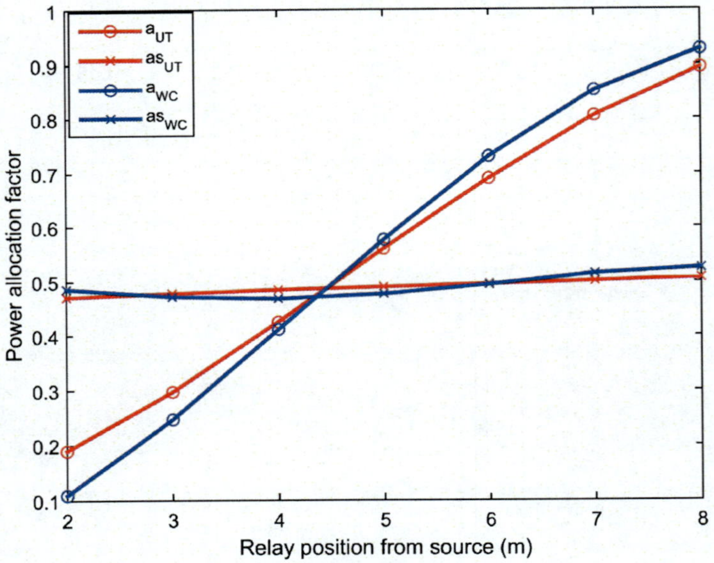

Fig. 6. Power allocation factors versus source-relay distance

Since optimal powers are allocated to the transmitting nodes in OPA, the proposed method shows better performance compared with other methods. Also, the secrecy performance is better for relays at the center of the network model. i.e., maximum secrecy is attained when the relay is at 6 m from the source. Since source and relay nodes take equal powers, EPA shows good performance only for the case when the relay is at the center of the model; i.e., at 5 m from the source. Its performance reduces for other relay positions and this is clearly understood from the Fig. 5. The impact of power allocation factors can be seen from Fig. 6. When the source-relay distance increases more source power is needed; hence a increases with distance. The eavesdropper near to source is considered here for the analysis. Since the eavesdropper position does not change, the jamming signal power $(1-a_s)$ remains almost constant at around 0.5.

One way to determine the complexity of the proposed algorithm is to analyze the average number of iterations needed for function convergence during simulation. The complexity analysis among the symmetric and asymmetric untrusted and worst case scenarios are presented in Table 2. The symmetric case requires less number of iterations for convergence than asymmetric case. Being the worst case from the viewpoint of secrecy, WC scenario requires more number of iterations to converge. It is evident that increase in SNR decreases the average number of iterations; showing less variation beyond 10 dB in all cases. It is also observed that the number of iterations increases with complexity; so also the memory usage and computational time.

Table 2. Complexity analysis among the proposed schemes

SNR(dB)	Average number of iterations			
	Symmetric case		Asymmetric case	
	UT	WC	UT	WC
0	61	61	65	81
2	63	63	59	79
4	49	55	51	75
6	43	48	49	67
8	40	45	47	51
10	39	43	41	53
12	38	39	43	51
14	38	39	41	53
16	39	40	43	49
18	40	41	45	53
20	40	41	43	51

6 Conclusions

In this paper, a source-based-jamming scheme that employs a gradient-free power optimization strategy named Nelder-Mead algorithm is formulated for secrecy enhancement in two-hop AF relaying networks with multiple untrusted relays and an external passive eavesdropper. Optimization methods such as two-dimensional exhaustive search and gradient-based methods are derived for comparison. The performance comparison of the proposed untrusted relaying scheme with worst case untrusted scenario and also with that of jamming scheme without power optimization (EPA strategy) is done. The complexity analysis is also performed. Simulation results show the performance improvement of proposed scheme over other optimization methods, EPA strategy and worst case jamming scenario.

References

1. Laneman, J.N., Tse, D.N.C., Wornell, G.W.: Cooperative diversity in wireless networks: efficient protocols and outage behavior. IEEE Trans. Inf. Theory **50**(12), 3062–3080 (2004)
2. Wang, W., Teh, K.C., Li, K.H.: Relay selection for secure successive AF relaying networks with untrusted nodes. IEEE Trans. Inf. Forensics Secur. **11**(11), 2466–2476 (2016)
3. He, X., Yener, A.: Cooperation with an untrusted relay: a secrecy perspective. IEEE Trans. Inf. Theory **56**(8), 3807–3827 (2010)
4. Wang, D., Bai, B., Zhao, W., Han, Z.: A Survey of optimization approaches for Wireless Physical Layer Security. IEEE Commun. Surv. Tutorials (2018). https://doi.org/10.1109/COMST.2018.2883144
5. Shemi, P.M., Jibukumar, M.G., Ali, M.A.: Enhancing secrecy in cooperative networks via power optimized source based jamming. In: Proceedings of the 2nd IEEE Middle East and North Africa Communications Conference (2019)
6. Wang, L., Elkashlan, M., Huang, J., Tran, N.H., Duong, T.Q.: Secure transmission with optimal power allocation in untrusted relay networks. IEEE Wireless Commun. Lett. **3**(3), 289–292 (2014)
7. Kuhestani, A., Mohammadi, A., Mohammadi, M.: Joint relay selection and power allocation in large-scale MIMO systems with untrusted relays and passive eavesdroppers. In: IEEE Transactions on Information Forensics and Security, pp. 341–355 (2018)
8. Lv, L., Chen, J., Yang, L., Kuo, Y.: Improving physical layer security in untrusted relay networks: cooperative jamming and power allocation. IET Commun. **11**(3), 393–399 (2017)
9. Mohammadi, A., Kuhestani, A.: Destination-based cooperative jamming in untrusted amplify-and-forward relay networks: resource allocation and performance study. IET Commun. **10**(1), 17–23 (2016)
10. Shemi, P.M., Jibukumar, M.G., Ali, M.A.: Nelder-Mead–based power optimization for secrecy enhancement in amplify-and-forward cooperative relay networks. Int. J. Commun. Syst. **32**(11), 532 (2019)
11. Nelder, J.A., Mead, R.: A simplex method for function minimization. Comput. J. Oxford University Press **7**(4), 308–313 (1965). https://doi.org/10.1093/comjnl/7.4.308
12. Sun, L., Zhang, T., Li, Y., Niu, H.: Performance study of two-hop amplify-and- forward systems with untrustworthy relay nodes. IEEE Trans. Vehicular Technol. **61**(8), 3801–3807 (2012)
13. Mathews, J.H., Fink, K.D.: Numerical methods using MATLAB Pearson (2004)
14. Gao, F., Han, H.: Implementing the Nelder-Mead simplex algorithm with adaptive parameters. Computational Optimization and Applications (2010)

Cyber Security Attacks on Identity and Location of Vehicle Ad-Hoc Networks

Haitham Alfehaid and Salim El Khrdiri(✉)

Department of Information Technology, College of Computer Qassim University, Buraydah, Saudi Arabia
{391100340,s.elkhediri}@qu.edu.sa

Abstract. Vehicle ad hoc network (VANET) technology arose from the mobile ad-hoc network (MANET) and the first mention of the term was in 2003. VANET allows vehicles to communicate with other vehicles and other intelligent transport systems (ITS) facilities in their location range. This paper presents a literature review and identify the security requirements needed to achieve a secure VANET environment. It also identifies a VANET cybersecurity attack, with a focus on attacks that target the location and identity of other vehicles as well as the ability to detect previous attacks. Moreover, this research started by introducing VANET architecture, communication and applications. Then, defining the important security requirements based on the three-information security triangle with an emphasis on the VANET environment. For this paper, we have developed an application of misbehaving attacks on the positioning and identity of VANET vehicles and then used the outputs of the application to analyze the misbehaviors attacks. Which is important for road safety and the protection of human life.

Keywords: VANET · Misbehavior · Cybersecurity · Attack delectation

1 Introduction

Technology is applied in the military, industrial, and civilian sectors. As technology relies on machines to carry out the usual and unusual tasks, human mistakes are reduced, and efficiency and speed of implementation are increased. As a result, the development and use of network and communications systems have increased and data no longer depend on a wired network to transmit as in the past. Data can now be transferred in massive quantities and speeds and via different types of wired and wireless networks.

Many ITS technologies and applications allow vehicles to communicate among themselves and with other facilities such as navigation systems, traffic light systems (TLS), plate recognition systems (PLS) as well as advanced applications.

These applications are called vehicular ad hoc networks (VANET) and aim to increase the efficient use of roads, enhance traffic safety, and reduce road accidents. The term VANET was mentioned for the first time in 2003.

There are three main types of the wireless network (WN) as shown in Fig. 1, cellular network, wireless LAN (WLAN) and wireless ad hoc networks (WANETs). WANETs

B. Ghita and S. Shiaeles (Eds.): INC 2020, LNNS 180, pp. 207–223, 2021.
https://doi.org/10.1007/978-3-030-64758-2_15

can be further divided into four applications; mobile wireless sensor network (MWSN), wireless sensor network (WSN), wireless mesh network (WMN) and mobile ad-hoc network MANETs [3].

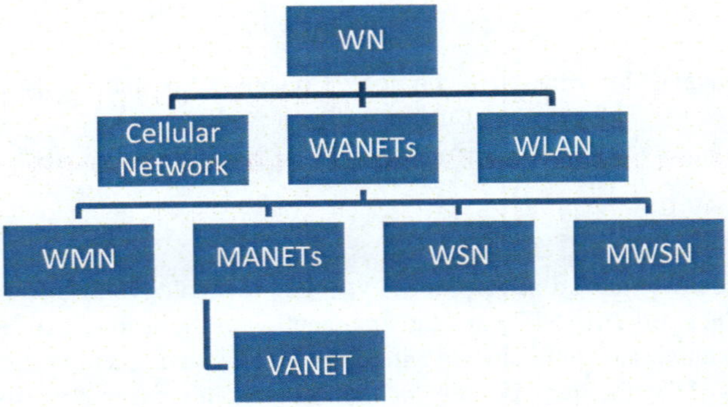

Fig. 1. Wireless networks type

VANET is made of decentralized networks that allow vehicles to communicate wirelessly among themselves and between other ITS facilities. VANET is a self-organized infrastructure that communicates without pre-planning. Vehicles act as a router and they broadcast all messages received to the vehicles and other infrastructure within their network range.

1.1 VANET Architecture

VANET architecture has three essential components; Onboard Unit (OBU) Roadside Unit (RSU) and Trusted Authority (TA) [5], as described below. The interactions of these components are shown in Fig. 2.

Onboard Unit (OBU): Every vehicle is embedded with this component, which enables vehicles to communicate with other vehicles. OBU acts like a transfer to communicate with RSU and other vehicles for safety information exchange. It has three further parts, i.e., a resource command processor, network device and sensors. Furthermore, it has a GPS device for gathering information about nearby vehicles.

Roadside Unit (RSU): The RSUs are the base stations installed at the corners of the road on the equal distance that aims to enable V2I communication. It has the feature of being able to communicate with the vehicle and other RSUs. The purpose of RSUs to prolong the communication range of the VANET. Furthermore, it also disseminates information to the ongoing vehicles on a highway and can enable connectivity to the cloud [6].

Trusted Authority (TA): The purpose of the TA is to provide secure connectivity to all vehicles on the RSU. It coordinates with all components of the VANET, aiming to avoid any type of threat and attack [6].

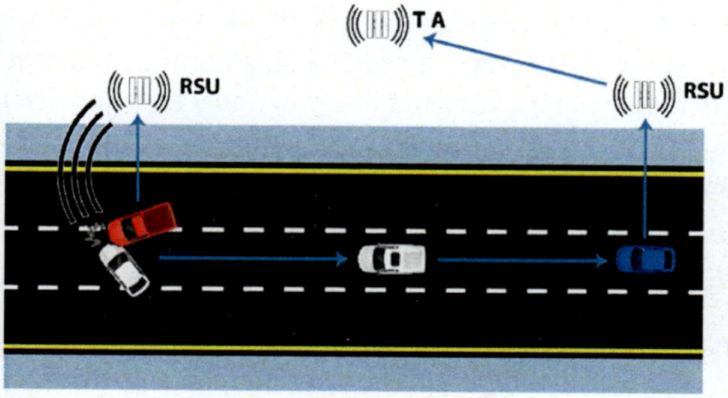

Fig. 2. VANET architecture

1.2 VANET Communication

The various communications are shown in Fig. 3, VANET has two-channel communications groups.

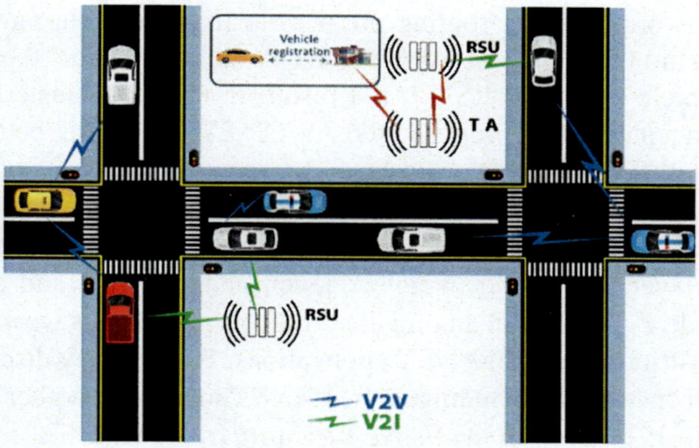

Fig. 3. VANET communications

The first type of communication is in-vehicle communication, where the user computing platform manages all the vehicle components such as radars, positioning systems and communication facilities.

The second type is a vehicle to vehicle communication (V2V) [8], which allows vehicles to connect and share information about speed and location wirelessly. V2V is used to help prevent accidents and increase road quality. The third type is a vehicle to infrastructure communication (V2I) [9], which aims to share data between vehicles and road infrastructure, including RSUs, traffic lights and car parking detectors.

1.3 Methodology

We developed an application to simulate and conduct the misbehaving attacks; after that, the application extracts the required dataset that contains the details of the vehicle as location, speed, and type. This research used the WEKA machine learning tool to conduct the experimental study. Moreover, we propose an algorithm to detect misbehaver attack in term of speed, identity and location. Finally, we analyzed and discussed the results of the dataset based on machine learning classification, which are Naïve Bayes, Random tree, Bagging and LogitBoost.

1.4 Literature Review

First, to understand what VANET is, we conducted a review of high-quality VANET research (i.e., papers with a high number of citations and from journals with a high impact factor). The articles explain comprehensive categories of the architecture, layers, applications, services and tools of VANET. Papers in this work were published between 2007–2017 and identified by searching ISI Scopus and Google Scholar.

S. Valayapalayam Kittusamy et al. [14] propose a new algorithm that organizes structure and cluster head (CH) of VANET appropriately. Moreover, they offer an adaptive clustering protocol (AWCP), which groups random nodes and then optimizes network parameters to accomplish an ideal CH. Besides, they presented a novel algorithm known as an "enhanced whale optimization algorithm" (EWOA) in a trusted clustering model, the vehicle network mobility routing protocol analyzes vehicle movement, identified speed, and position. Mainly, the AWCP-EWOA model explains the distance between the trusted vehicle node and RSU. As a result, in terms of clustering efficiency and mobility enhancement, they proposed the AWCP-EWOA model betters in comparison with AWCP and AWCP-Whale protocols.

The challenges of VANET communication and applications have been analyzed by F. Cunha et al. [15]. They also reviewed various types of network protocol stacks, such as a physical layer, Mac layer, Network layer, and transport and application layers. Then, they made a comparison among the protocols in previous work in their paper and discussed the various type of VANET applications. Finally, they discussed the research problem to will open doors to finding a new VANET solution. Another survey paper by H. Hasrouny et al. [19] discussed the VANET security framework. The authors provided an overview of security challenges and requirements of VANET and they gave facts of the well-known protocols and the new security architectures. Then, they made a comparison of the solutions mentioned above. In the end, they opened a discussion topic which can help researchers in the future.

2 Cyber Security Attack on Identity and Positioning

This section focuses on four kinds of attacks that interrupt vehicle positioning location and identity. Thus, these types of attacks may lead to a slowing down of vehicle speed and effect on road safety and quality. Next section will present the misbehaviors attacks on the identity and location of vehicles, which may lead to reduce road safety and affect the comfort of drivers [25, 26].

2.1 Single-Vehicle Faked Location (SVFL)

A single-vehicle faked location (SVFL) attack happens when the attacker duplicates its identity (ID) to be another vehicle, acting as a virtual vehicle that appears in the street. In this type of attack, the attacker vehicle broadcasts its identity by broadcast messages by using the BMS message but in a fake location, to be shown to other vehicles as a real vehicle in the same street path direction. However, the vehicle attacker broadcasts a random position located at different times intervals for the fake position locations.

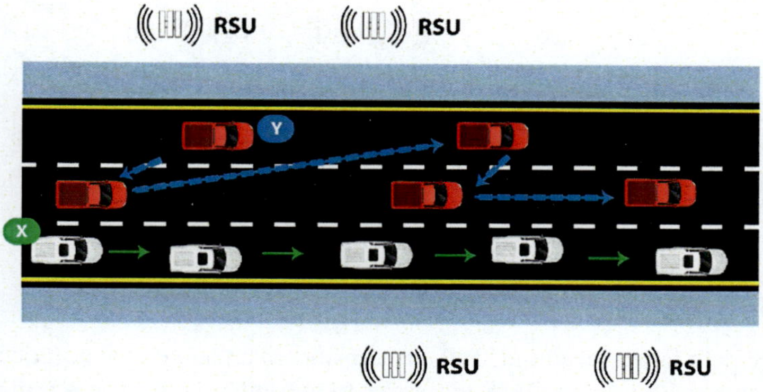

Fig. 4. Single-vehicle faked location illustrate

Figure 4 illustrates two vehicles; X and Y. Vehicle X is a real vehicle, shown as the white vehicle linked with the solid arrow, which presents the actual movement path at different time intervals. Vehicle X broadcasts BMS messages to show itself as another vehicle to others. A virtual vehicle, in this example vehicle Y, similarly broadcasts BMS messages to others. Vehicle Y is the red vehicle and linked with the dotted arrow. Vehicle X broadcasts random position locations for vehicle Y in the same street direction, which presents the random virtual movement path at different time intervals.

2.2 Multi-vehicle Faked Location (MVFL)

The multi-vehicle faked location (MVFL) occurs when the attacker broadcasts messages using multiple fake IDs by using another vehicle ID or by fabricating one, to present to others as a real vehicle. The attacker also positions the location of the virtual vehicles randomly in the same street direction. Moreover, the virtual vehicles use one location at the same time for each and do not use it again.

Figure 5 illustrates three vehicles X, Y, and Z. Vehicle X is a real vehicle that is shown as the white vehicle linked together with a solid arrow that gives the real path of the vehicle at different time intervals. Vehicle X also produces virtual vehicles by broadcasting messages using its fabricated IDs. Vehicles Y and Z are shown as red and white vehicles, respectively, and are virtual vehicles. These vehicles are linked together with a dotted arrow, which presents the virtual movement path at different time intervals. Vehicle X generates the IDs of Y and Z by using another vehicle's ID or by fabricating

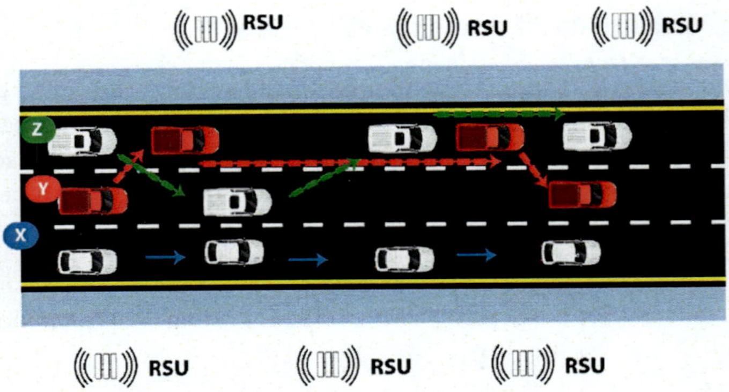

Fig. 5. Multi-vehicle faked location illustrate

an ID. The location of vehicles Y and Z are positioned randomly in the same street direction path by vehicle X.

2.3 Single-Path Faked Location (SPFL)

The single path faked location (SPFL) is considered an acute attack. It occurs when the vehicle attacker broadcasts BMS messages to presents to others as a real vehicle when it is a virtual vehicle reincarnating its ID. The positioning location of this type of attack is predefined to be like a usual movement.

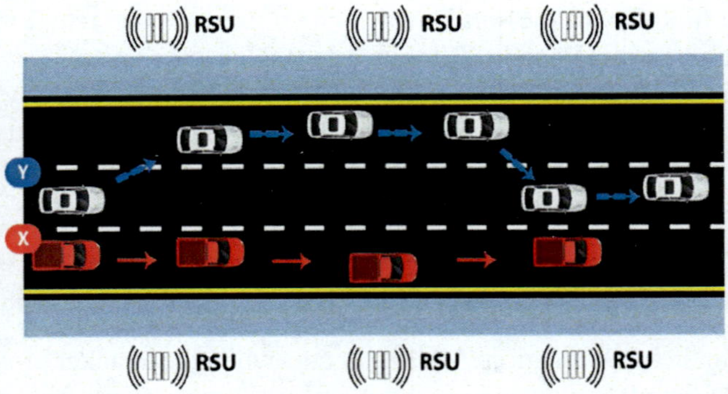

Fig. 6. Single path faked location illustrate

Figure 6 shows two vehicles, X and Y. Vehicle X is a real vehicle and shown as the red vehicle linked together with the solid arrow, which presents the actual movement path at different time intervals. Vehicle Y, however, is a virtual vehicle and shown as the white vehicle linked together with the dotted arrow, which presents the virtual movement path at different time intervals. In this example, vehicle X has duplicated its ID and the positioning location for vehicle Y is predefined.

2.4 Multi-path Faked Location (MPFL)

The multi-path faked location (MPFL) attack is almost the same as the SPFL and is similarly considered an acute attack that cannot be detected easily. Moreover, MPFL leads to issues in traffic safety because this type of attack causes fake street traffic, which disrupts traffic and time. An MPFL attack happens when a vehicle attacker establishes and broadcasts multiple virtual vehicles at the same time. The location positioning of virtual vehicles is predefined by the attacker and they do not use location twice.

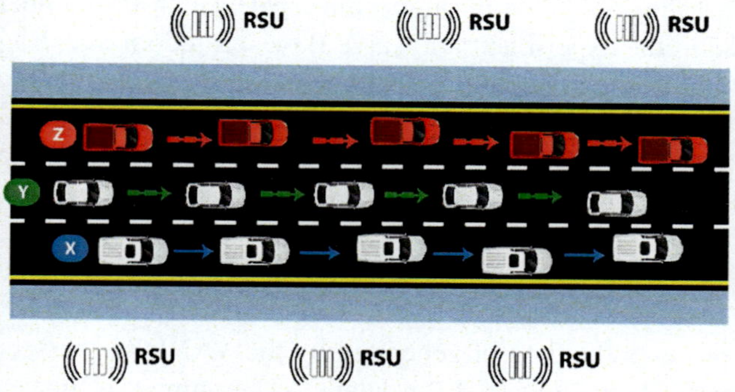

Fig. 7. Multi-path faked location illustrate

Figure 7 shows three vehicles, X, Y and Z. Vehicle X is the attacker vehicle, shown as the white vehicle and linked together with the solid arrow, which presents the actual path at different time intervals.

Vehicles Y and Z are virtually created by vehicle X and are shown the white and red vehicles, respectively, and are linked together with the dotted arrow that presents the virtual path at different time intervals. Vehicle X fakes the identity of vehicles Y and Z, with the positioning location of vehicle Y is predefined.

3 Cybersecurity Attack Experiment

This section discusses the research tool that used in this study. The configuration of the simulation and the impact of the attack also present, the generated dataset from the VANET simulation, the experiment on the dataset and work with a machine learning tool to analyze the dataset.

3.1 Configuration of the Experiment on the Simulation

This section details the configurations used in the experiment simulation. The road setup was a highway of 10 km length and two road directions with multiple tracks in each. The number of vehicles simulated in the experiment was 12,000 and the random speed was set between 10 km/h to 180 km/h.

3.2 The Machine Learning Software

This study was conducted using the Waikato Environment for Knowledge Analysis (WEKA) machine learning tool [27]. WEKA is easy to use, and you do not need programming skills to use it. In addition, WEKA has an existing library for the developer to use.

3.3 Classification Algorithms

The machine learning classification algorithm can deal with any number of instances. The classifiers used in the experiment are Naïve Bayes, Random tree, Bagging, LogitBoost and J48 [28, 29]. The classification is completed in two steps: first, create a model by train on a group of instances by applying a classification algorithm. Second, Measure the performance and accuracy of the trained data by testing the extracted model, with ten folds cross-validation as an option.

3.4 The Generated Dataset

This section presents the dataset generated by the VANET cybersecurity attack simulation program [30] (see Sect. 4.3 for details). The number of instances of the dataset generated by simulation was 263,889. Furthermore, the number of attributes was 7 (Id, location (x, y), speed (x, y), type, attack type). The timestamp and steps were excluded the avoid overfitting. Moreover, the dataset is labeled (multi-class) and contains five classes of attack (None, SVFL, MVFL, SPFL, MPFL).

4 Cybersecurity Attack Results

The goal of this section is to discuss the experiment on the attack that described in Sect. 2. Then, presenting the machine learning tools, classification, and the generated dataset used in this study.

4.1 Naïve Bayes Classifier Results

Figure 8 present the first classifier Naïve Bayes [31]. The five parameters that will be discussed are the TP rate, FP rate, Precision, Recall and F-measure.

The None attack category achieved the highest ratio in the TP rate at 100, the SVFL attack achieved 3.1% which is the lowest ratio. The average ratio of the TP rate was 46.1%. The FP, the MPFL attack achieved a ratio of 19.2%. The average FP rate was 19.2%.

The None class achieved the highest rate at 100% in Precision, the SVFL attack was the lowest rate at 25%. The average precision rate was 42.2%.

The None class achieved the highest ratio at 100% in Recall. The average ratio of recall was 46.1%, which equates to the average TP rate.

Finally, The None class achieved the highest ratio at 100% in F-measure. The SPFL attack achieved 5.5%. The f-measure average rate was 39.9%.

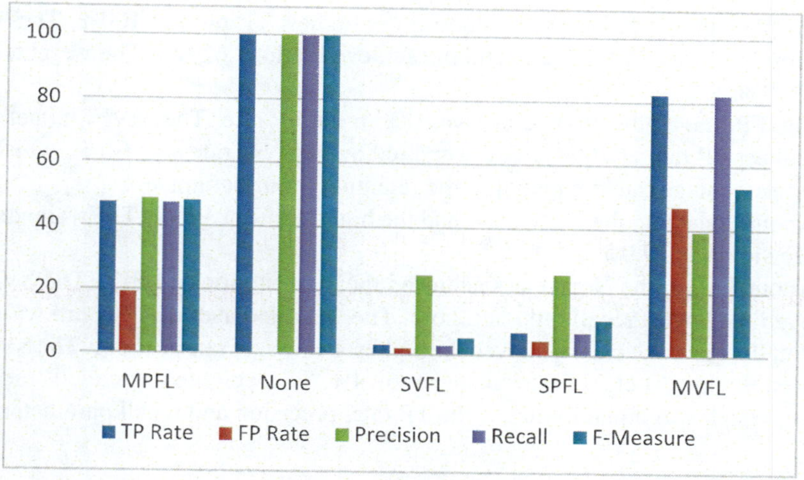

Fig. 8. Naïve Bayes chart

4.2 Random Tree Classifier Results

This section present results for the random tree classifier [32] analyses as shown in Fig. 9. The five parameters of performance were the TP Rate, FP Rate, Precision, Recall, and F-measure.

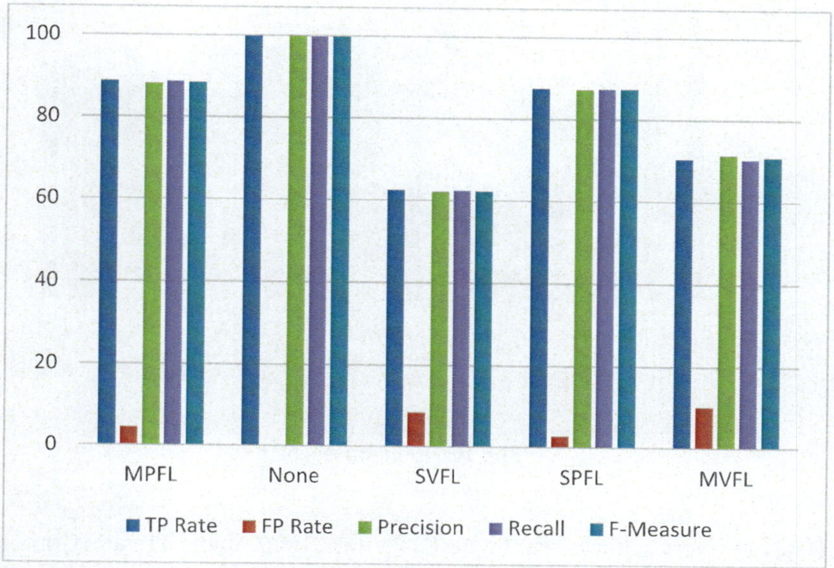

Fig. 9. Random tree chart

In TP ratio, the None class achieved the highest TP ratio at 100%. The lowest TP ratio reported was SVFL attack and the achieved ratio at 62.6%. The weighted average was 79.9%.

The FP rate of the None class was 0% in the FP rate. The MVFL attack achieved the highest FP rate at 10.3%. The weighted average FP rate was 6.1%, which was the lowest percentage rate compared to the remaining four parameters.

About Precision, the None class had the highest rate at 100%. The weighted average of precision was 79.9%.

About Recall, the None class achieved the highest ratio at 100%. The SVFL attack achieved the lowest recall ratio at 62.6%. The weighted average of recall was 79.9%.

Finally, the None class achieved the highest f-measure rate at 100%. The SVFL attack reported the lowest class performance at 62.4%. The weighted average F-measure was 79.9%, which was equal to that of the TP rate, precision and recall parameters.

4.3 LogitBoost Classifier Results

As shown in Fig. 10, the results of the five classes of attack for LogitBoost classifier analyses [33]. The five parameters of performance are the TP rate, FP rate, Precision, Recall and F-measure.

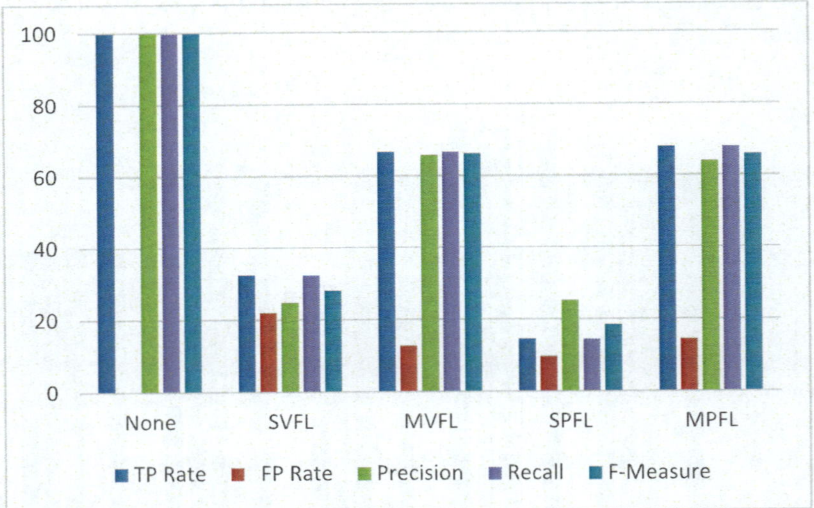

Fig. 10. Logit boost chart

The None class achieved the highest TP rate at 100% in the TP rate. Conversely, the SVFL attack was the lowest-rated class performance at 4.5%. The weighted average TP rate was 52.6%.

In the FP rate, all classifications were less than the MPFL attack, which rated at 33.4%. The None class is the lowest performance ratio achieved and rated 0%. The weighted average FP rate was 16.7%, which was the lowest ratio of all five parameters.

About Precision, The None class achieved the highest precision rate at 100%. The weighted average for precision was 46.2%.

In the Recall, the SVFL attack reported the lowest percentage of the recall at 4.5%. In contrast, the None class achieved the highest rate at 100%. The weighted average of recall was equal to that of the TP rate at 52.6%.

Finally, in F-measure the None class achieved the highest rate at 100%. The weighted average f-measure rate was 46%.

4.4 Bagging Classifier Results

Figure 11 present the analytical results for the bagging classifier [34]. The five parameters of performance were the TP Rate, FP Rate, Precision, Recall, and F-measure.

The None class achieved the highest TP ratio at 100% in the TP rate. The lowest percentage reported was the SVFL attack at 64.3%. The weighted average TP rate was 87.3%.

In the FP rate the FP rate of None class was 0%. The MVFL attack achieved the highest FP rate at 13.2%. The weighted average FP rate was 4.3%, which was the lowest weighted average in comparison with the other four parameters.

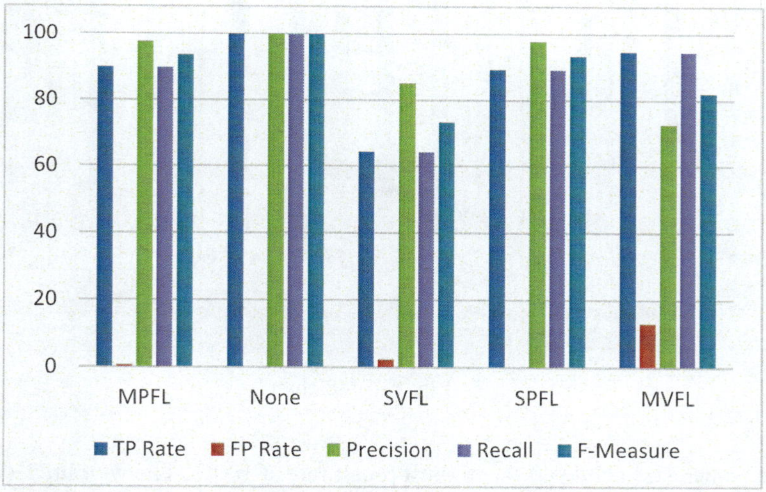

Fig. 11. Bagging chart

The None class as the highest ratio at 100% in the Precision. The MPFL attack was lower in performance than the None class at 97.6%. The weighted average of precision was 88.7%, which was the highest of all five parameters.

The None class achieved the highest ratio at 100% in the Recall. The SVFL attack achieved the lowest recall ratio at 64.3%. The weighted average of recall was 87.3%.

Finally, The None class achieved the highest f-measure rate at 100% in the F-measure.

The SVFL attack reported the lowest class performance at 73.3%. The weighted average f-measure was 87.3%, which was equal to the TP rate and recall parameter weighted averages.

4.5 J48 Classifier Result

Figure 12 shows the results for the five classes of attack using the J48 classifier [41]. The five parameters of performance were the TP Rate, FP Rate, Precision, Recall, and F-measure.

The None class achieved the highest TP rate at 100%. The SVFL attack achieved a lower classification performance at 63.3%. The weighted average TP rate was 87.5%.

The SPFL attacks achieved an almost lower FP rate at 1.2%. The None class reported a lower classification performance at 0%. The weighted average FP rate was 4.3%, which was the lowest ratio among all parameters.

The None class achieved the highest precision ratio at 100%. The MVFL attack reported a lower classification performance at 74.6%. The weighted average of precision was 88.7%, which was the highest ratio among all parameters.

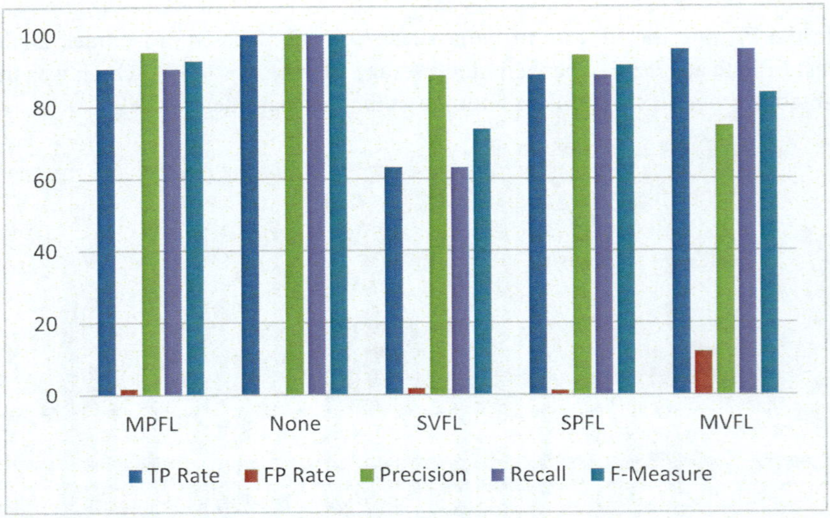

Fig. 12. J48 chart

The None class achieved the highest recall rate at 100%. The weighted average ratio of recall was 87.5%, which equates to the TP rate.

The highest rate of F-measure was for None class at 100%. The weighted average f-measure was 87.3%.

5 Discussion and Results

This section will discuss the experiment result on term of time and accuracy. Then, presenting the proposed algorithm that detect the misbehavior attack on identity, speed and position location.

5.1 Time and Accuracy Performance

The time taken to build a model/sec is an essential factor in choosing the best classifier according to accuracy and less time needed to build a model [37].

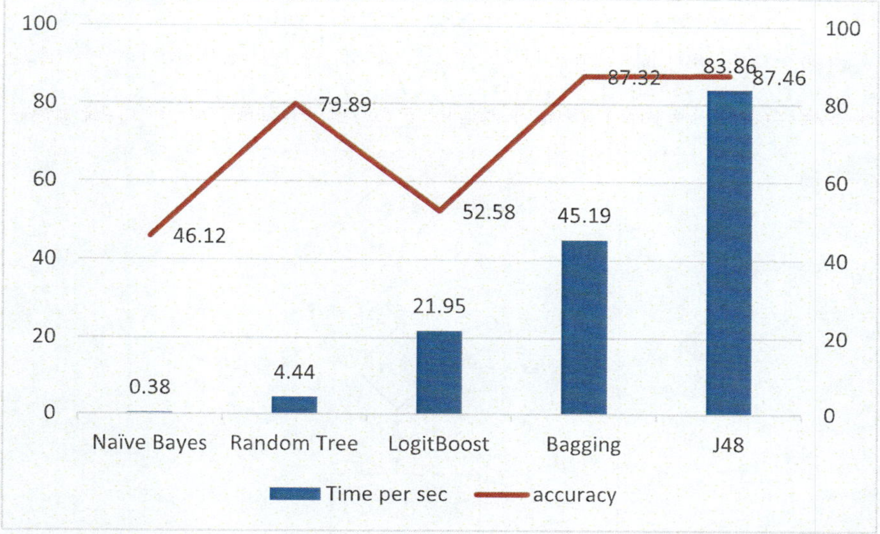

Fig. 13. Time taken to build a model/sec and accuracy

Figure 13 presents the detailed analytical results for the time taken to build a model/sec related to the accuracy of five classifiers, Naïve Bayes, Random Tree, LogitBoost, Bagging, and J48 classifiers. The parameters of performance were time and accuracy.

The Random Tree classifier achieved the best performance at 4.44 s with an accuracy of 79.89%. The LogitBoost and Naïve Bayes classifiers had a lower performance at 21.95 s and 52.58% accuracy and 0.38 s and 46.12%, respectively. The J48 classifier had a time of 83.86 s and an accuracy of 83.46%, which was the best accuracy compared with the other classifiers. Finally, the time of the Bagging classifier to build model was 45.19 s and the accuracy was 87.32%.

5.2 VANET Attack Detection Flowchart

The following section will present the flowchart of attack detection as shown in Fig. 14.

5.3 The Generated Dataset

This section presents the dataset generated by the simulation program and analysis of the dataset. The number of instances of the dataset generated by simulation was 263,889. Furthermore, the number of attributes was eleven (Id, location x and y, speed x and y, type, attack type, detector). The timestamp, steps, and detection reason were excluded the avoid overfitting. In addition, the dataset is labeled (two-class) and contains two classes (malicious, normal vehicle).

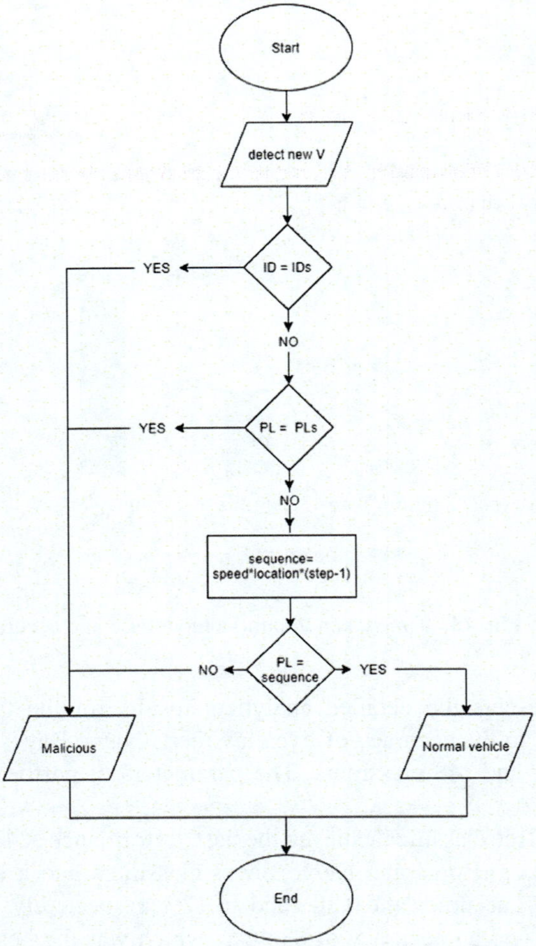

Fig. 14. Attack detecting algorithm flowchart

5.4 Detection's Time and Accuracy Performance

The time taken to build a model/sec related to accuracy is an essential factor in choosing the best classifier. The ideal model will have high accuracy and take less time needed to build.

Figure 15 presents the detailed analytical results for the time taken to build a model/sec related to the accuracy of the five classifiers (Naïve Bayes, Random Tree, LogitBoost, Bagging and J48). The parameters of performance were time and accuracy.

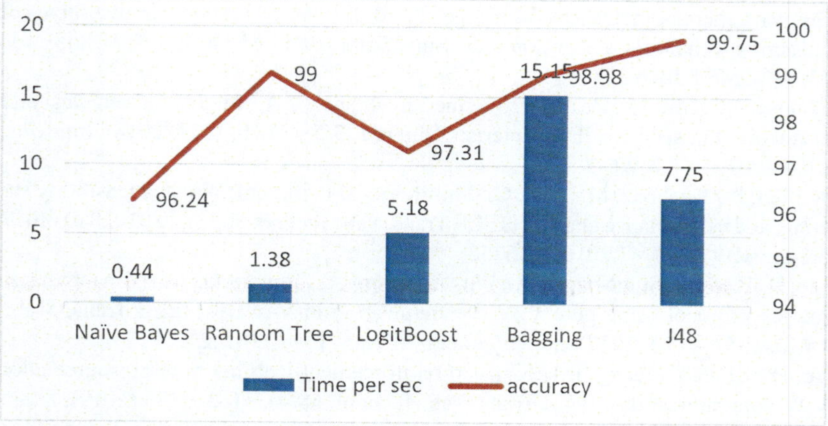

Fig. 15. The detection's time taken to build a model/sec and accuracy chart

The Random Tree classifier achieved the best performance at 1.38 s with an accuracy of 99%. The Bagging classifiers reported the lowest classifier performance at 15.15 s, with an accuracy of 98.98%.

The J48 classifier reported time at 7.75 s and accuracy at 99.75%, which was the best accuracy among all other classifiers. Finally, the time it too the Naïve Bayes and LogitBoost classifiers to build models was 0.44 s and 5.18, with accuracy at 96.24% and 97.31%, respectively.

6 Conclusion and Future Work

This study has presented a thorough literature review of high quality and high impact papers in the VANET field papers. This research review the security requirements that should be in all electronic systems, including VANET, based on the information security triangle (confidentiality, integrity, and availability) and proposed a TA for authorized vehicles and techniques that can be added to improve the accuracy of vehicle validation.

In the experiment, Fake BMS messages sent by attackers represented misbehavior attacks in the VANET environment. Moreover, our research focused on four types of attacks, SVFL, MVFL, SPFL and MPFL. Identifying various types of attacks on identity and positioning locations that face VANET has impacted the security, safety and comfort of drivers.

References

1. Road Traffic Injuries. https://www.who.int/news-room/fact-sheets/detail/road-traffic-injuries. Accessed 08 Feb 2020
2. Traffic Accident Statistics as of 1439 H - Datasets - Saudi Open Data. https://data.gov.sa/Data/en/dataset/traffic-accident-statistics-as-of-1439-h. Accessed 08 Feb 2020
3. Glass, S., Mahgoub, I., Rathod, M.: Leveraging MANET-based cooperative cache discovery techniques in VANETs: a survey and analysis. IEEE Commun. Surv. Tutor. **19**(4), 2640–2661 (2017). https://doi.org/10.1109/COMST.2017.2707926

4. Cooper, C., Franklin, D., Ros, M., Safaei, F., Abolhasan, M.: A comparative survey of VANET clustering techniques. IEEE Commun. Surv. Tutor. **19**(1), 657–681 (2017). https://doi.org/10.1109/COMST.2016.2611524
5. Contreras-Castillo, J., Zeadally, S., Guerrero-Ibanez, J.A.: Internet of vehicles: architecture, protocols, and security. IEEE Internet Things J. **5**(5), 3701–3709 (2018). https://doi.org/10.1109/JIOT.2017.2690902
6. Atallah, R., Khabbaz, M., Assi, C.: Multihop V2I communications: a feasibility study, modeling, and performance analysis. IEEE Trans. Veh. Technol. **66**(3), 2801–2810 (2017). https://doi.org/10.1109/TVT.2016.2586758
7. Lin, D., Kang, J., Squicciarini, A., Wu, Y., Gurung, S., Tonguz, O.: MoZo: a moving zone based routing protocol using pure V2V communication in VANETs. IEEE Trans. Mob. Comput. **16**(5), 1357–1370 (2017). https://doi.org/10.1109/TMC.2016.2592915
8. Ye, H., Li, G.Y., Juang, B.H.F.: Deep reinforcement learning based resource allocation for V2V communications. IEEE Trans. Veh. Technol. **68**(4), 3163–3173 (2019). https://doi.org/10.1109/TVT.2019.2897134
9. Dey, K.C., Rayamajhi, A., Chowdhury, M., Bhavsar, P., Martin, J.: Vehicle-to-vehicle (V2V) and vehicle-to-infrastructure (V2I) communication in a heterogeneous wireless network - Performance evaluation. Transp. Res. Part C Emerg. Technol. **68**, 168–184 (2016). https://doi.org/10.1016/j.trc.2016.03.008
10. Combs, T.S., Sandt, L.S., Clamann, M.P., McDonald, N.C.: Automated vehicles and pedestrian safety: exploring the promise and limits of pedestrian detection. Am. J. Prev. Med. **56**(1), 1–7 (2019). https://doi.org/10.1016/j.amepre.2018.06.024
11. Yang, T., Zhang, Y., Tan, J., Qiu, T.Z.: Research on forward collision warning system based on connected vehicle V2V communication. In: 2019 5th International Conference on Transportation Information and Safety (ICTIS), pp. 1174–1181 (2019). https://doi.org/10.1109/ICTIS.2019.8883534
12. Oliveira, R., Montez, C., Boukerche, A., Wangham, M.S.: Reliable data dissemination protocol for VANET traffic safety applications. Ad Hoc Netw. **63**, 30–44 (2017). https://doi.org/10.1016/j.adhoc.2017.05.002
13. Cavalcanti, E.R., Rodrigues de Souza, J.A., Spohn, M.A., De Morais Gomes, R.C., Ferreira Da Costa, A.F.B.: VANETs' research over the past decade: overview, credibility, and trends. Comput. Commun. Rev.**48**(2), 31–39 (2018). https://doi.org/10.1145/3213232.3213237
14. Valayapalayam Kittusamy, S.R., Elhoseny, M., Kathiresan, S.: An enhanced whale optimization algorithm for vehicular communication networks. Int. J. Commun. Syst. e3953 (2019). https://doi.org/10.1002/dac.3953
15. Cunha, F., et al.: Data communication in VANETs: protocols, applications and challenges. Ad Hoc Netw. **44**, 90–103 (2016). https://doi.org/10.1016/j.adhoc.2016.02.017
16. Boussoufa-Lahlah, S., Semchedine, F., Bouallouche-Medjkoune, L.: Geographic routing protocols for Vehicular Ad hoc NETworks (VANETs): a survey. Veh. Commun. **11**, 20–31 (2018). https://doi.org/10.1016/j.vehcom.2018.01.006
17. Sanguesa, J.A., Fogue, M., Garrido, P., Martinez, F.J., Cano, J.C., Calafate, C.T.: A survey and comparative study of broadcast warning message dissemination schemes for VANETs. Mob. Inf. Syst. **2016** (2016). https://doi.org/10.1155/2016/8714142
18. Engoulou, R.G., Bellaïche, M., Pierre, S., Quintero, A.: VANET security surveys. Comput. Commun. **44**, 1–3 (2014). https://doi.org/10.1016/j.comcom.2014.02.020
19. Hasrouny, H., Samhat, A.E., Bassil, C., Laouiti, A.: VANet security challenges and solutions: a survey. Veh. Commun. **7**, 7–20 (2017). https://doi.org/10.1016/j.vehcom.2017.01.002
20. Sumra, I.A., Bin Hasbullah, H., Bin AbManan, J.L.: Attacks on security goals (confidentiality, integrity, availability) in VANET: a survey. Adv. Intell. Syst. Comput. **306**, 51–61 (2015). https://doi.org/10.1007/978-981-287-158-9_5

21. Wan, C., Zhang, J.: Efficient identity-based data transmission for VANET. J. Ambient Intell. Humaniz. Comput. **9**(6), 1861–1871 (2018). https://doi.org/10.1007/s12652-017-0650-x
22. Zhang, L., Wu, Q., Domingo-Ferrer, J., Qin, B., Hu, C.: Distributed aggregate privacy-preserving authentication in VANETs. IEEE Trans. Intell. Transp. Syst. **18**(3), 516–526 (2017). https://doi.org/10.1109/TITS.2016.2579162
23. Tzeng, S.F., Horng, S.J., Li, T., Wang, X., Huang, P.H., Khan, M.K.: Enhancing security and privacy for identity-based batch verification scheme in VANETs. IEEE Trans. Veh. Technol. **66**(4), 3235–3248 (2017). https://doi.org/10.1109/TVT.2015.2406877
24. Manvi, S.S., Tangade, S.: A survey on authentication schemes in VANETs for secured communication. Veh. Commun. **9**, 19–30 (2017). https://doi.org/10.1016/j.vehcom.2017.02.001
25. Gurumoorthi, E., Ayyasamy, A.: An intelligent fuzzy based location aided routing in vehicular ad hoc networks. Int. J. Innov. Technol. Explor. Eng. **8**(11), 1946–1955 (2019). https://doi.org/10.35940/ijitee.K2141.0981119
26. Zidani, F., Semchedine, F., Ayaida, M.: Estimation of Neighbors Position privacy scheme with an Adaptive Beaconing approach for location privacy in VANETs. Comput. Electr. Eng. **71**, 359–371 (2018). https://doi.org/10.1016/j.compeleceng.2018.07.040
27. Kotthoff, L., Thornton, C., Hutter, F.: User Guide for Auto-WEKA version 2.6 (2017)
28. Das, R., Khilar, P.M.: Driver behaviour profiling in VANETs: comparison of ensemble machine learning techniques. In: 2019 IEEE 1st International Conference on Energy, Systems and Information Processing, ICESIP 2019 (2019). https://doi.org/10.1109/ICESIP46348.2019.8938266
29. Nishani, L., Biba, M.: Machine learning for intrusion detection in MANET: a state-of-the-art survey. J. Intell. Inf. Syst. **46**(2), 391–407 (2016). https://doi.org/10.1007/s10844-015-0387-y
30. alfehaid/Vehical-Ad-hoc-Network. https://github.com/alfehaid/Vehical-Ad-hoc-Network. Accessed 04 Apr 2020
31. Saritas, M.M., Yasar, A.: Performance analysis of ANN and Naive Bayes classification algorithm for data classification. Int. J. Intell. Syst. Appl. Eng. **7**(2), 88–91 (2019). https://doi.org/10.18201//ijisae.2019252786
32. Kevric, J., Jukic, S., Subasi, A.: An effective combining classifier approach using tree algorithms for network intrusion detection. Neural Comput. Appl. **28**(1), 1051–1058 (2017). https://doi.org/10.1007/s00521-016-2418-1
33. Kamarudin, M.H., Maple, C., Watson, T., Safa, N.S.: A LogitBoost-based algorithm for detecting known and unknown web attacks. IEEE Access **5**, 26190–26200 (2017). https://doi.org/10.1109/ACCESS.2017.2766844
34. Tharwat, A., Gaber, T., Awad, Y.M., Dey, N., Hassanien, A.E.: Plants identification using feature fusion technique and bagging classifier. Adv. Intell. Syst. Comput. **407**, 461–471 (2016). https://doi.org/10.1007/978-3-319-26690-9_41

Challenges in Developing a Wireless Sensor Network for an Agricultural Monitoring and Decision System

Max Frohberg(✉), Stefan Weidling(✉), and Peter Langendoerfer(✉)

IHP – Leibniz-Institut für innovative Mikroelektronik, Frankfurt (Oder), Germany
{frohberg,weidling,langendoerfer}@ihp-microelectronics.com,
https://www.ihp-microelectronics.com/

Abstract. Demand for food, efficient use of resources and the need for climate change adaptation are conflicting objectives of today's agriculture. Wireless Sensor Networks (WSNs) could help to balance these contradicting requirements. A decisive advantage of a WSN is that data can be obtained from the sensors at any time without the physical presence of farmers. But in addition to a large number of technical challenges, a major challenge is to monitor necessary parameters with a sufficiently high temporal and spatial resolution. The present work discusses those challenges as a case study. Furthermore, an approach to designing a WSN for sensor-assisted landscape monitoring is proposed, that aims to support small-scale real time acquisition of site-specific requirements. Continuous monitoring is intended to lay the foundation for agricultural management strategies to be adapted at any time using real-time information.

Keywords: WSN · Wireless sensor network · Low-power · Smart farming

1 Introduction

Today, agricultural systems are torn between the increasing demand for agricultural products, the scarcity of resources, the reduction of biodiversity and climate change [1]. An increase in yield to ensure the need for food, which was forecast for 2050, needs to be achieved without exceeding the resilience limits of ecological systems [2–4]. A sustainable and resource-efficient increase in agricultural production is necessary. The apparently contradictory goals [1] could be achieved using smart farming, the use of modern information and communication technologies for the digitalisation of agriculture. New, highly complex smart farming systems (system of systems), could favor the development of ecological agriculture, which provides diverse ecosystem services and enables society

This work was supported by the Federal Ministry of Education and Research (BMBF) under research grant number 031B0729C.

B. Ghita and S. Shiaeles (Eds.): INC 2020, LNNS 180, pp. 224–240, 2021.
https://doi.org/10.1007/978-3-030-64758-2_16

to reward them [5]. WSNs could provide the data basis for such systems. WSN refers to a group of spatially dispersed and dedicated sensors for monitoring and recording the physical conditions of the environment. They can organize the collected data at a central location for further processing and decision making. Their use in an agriculture context could offer many advantages. A key advantage of a WSN is that data from the field can be retrieved from the sensors at any time without the physical presence of a farmer. An essential difference between traditional wired systems and WSNs is the dynamics of the network topology [6], the reliability of message transmission [7] and the limited energy budget [8]. While wired networks enable very reliable message transmission and external energy supply, this is not the case for WSNs. These constraints create numerous challenges when using WSNs in agriculture, such as determining an optimal distribution strategy, measurement periods, adaptation to variable climatic and soil conditions, energy efficiency and lifetime, costs, communication range, scalability, heterogeneity, security requirements, fault tolerance and reliability, temporal accuracy and synchronicity [9,10].

As part of the Digital Agricultural Knowledge and Information System (DAKIS) project[1], which intends to use advancing digitalization to integrate Ecosystem Services (ESS) and biodiversity into modern planning, production and marketing processes, a WSN has to be developed and implemented on experimental fields in two pilot regions in order to capture site-specific parameters in real time on a small scale. It is proposed to use a WSN to lay the foundation for adapting agricultural management strategies based on a Decision Support System (DSS) at any time using real-time information, as shown in Fig. 1. Another research question that arises in this context is whether this provides continuous feedback on the success of agricultural measures. The underlying models, DSS and agricultural measures are not in the scope of this paper. This paper discusses project-specific requirements, constraints and challenges that have to be considered while designing a WSN to support small-scale real-time acquisition of site-specific parameters for this project. This paper is organized as follows. The relevance of WSNs in the field of precision agriculture and environmental monitoring is shown in Sect. 2. Section 3 presents the requirements and constraints of such a WSN. The challenges to be overcome are discussed in detail in Sect. 4. The proposed approach for the design of a project specific WSN is introduced in Sect. 5. Conclusions are drawn in Sect. 6.

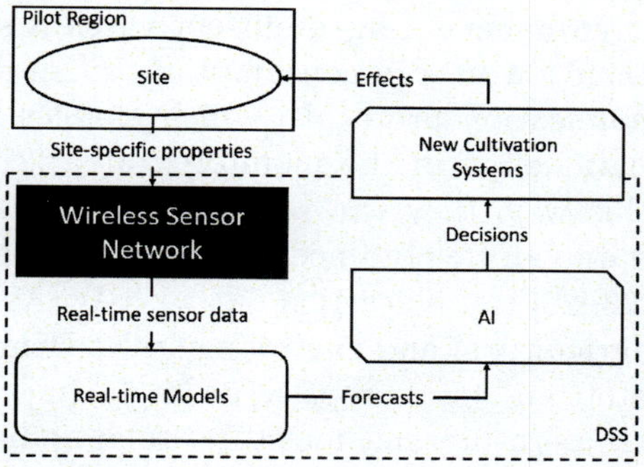

Fig. 1. WSN as part of a DSS

[1] Project website: https://adz-dakis.com/.

2 Relevance of WSNs in the Field of Agriculture Monitoring and Decision Systems

Numerous publications have already recognized the value of WSNs in precision agriculture and environmental monitoring [11–15]. Today WSNs are an integral part of it [9,10,16–22]. The use of wired networks is often impossible or impractical due to the special circumstances of the agricultural sector. The decisive advantage of a WSN is that data can be obtained from the sensors at any time without the need for the farmer to constantly visit the field [11]. Thanks to the rapid development in the field of WSNs and miniaturisation, precision agriculture was established [23]. To improve reliability and availability, the sensor networks provide mechanisms for self-organization, self-configuration as well as self-diagnosis and self-healing and allow flexible expansion [17]. Using heterogeneous sensor installations, agricultural issues can be mapped onto the sensor networks and targeted data acquisition can be carried out. Sensor networks are an important tool to measure and understand the complex coupled dynamics of biological systems [24]. They make it possible to increase efficiency, productivity and profitability while at the same time minimizing unintended effects on the environment [19]. Intelligent wireless sensor nodes can be distributed thanks to their low energy consumption, small dimensions and simple, high-density communication protocols, which enables precise and fine-grained measurement of parameters in the monitored area [22]. If critical parameters are measured in this way, they can be reacted to in a timely and localized manner, since the data can be obtained in real time [22]. For example, crop condition monitoring [23–29] can be carried out with WSNs. Through real-time monitoring of critical agricultural and environmental parameters, the farmer is able to react immediately and treat crops accordingly. Important parameters include, for example, air temperature and moisture, solar radiation, soil temperature, moisture and pH values, electrical conductivity, evapotranspiration, leaf wetness, carbon dioxide and nitrogen concentration as well as the detection of economically important pests. They are used for the irrigation and efficient use of water resources by monitoring soil moisture [30–32] and to monitor water quality [33], water distribution for leak detection [34], greenhouses [35,36] or greenhouse gases [37]. Precision viticulture can also benefit from them [38,39]. New opportunities in the use of WSNs in smart farming are offered, for example, by incorporating newer technical concepts such as sensor cloud computing, big data and the Internet of Things (IoT) [9].

3 Requirements and Constraints

Most requirements regarding hard- and software have their origin in basic WSN concepts with a direct influence on the design decisions. This is all about long-term operation with small and efficient battery-driven devices and the resulting low power operation. Further requirements are based on project goals, the support of partners (hard- and software interfaces, transport of additional sensor

Table 1. Main requirements and constraints

Level	Requirements	Contraints
Project (goals, partners)	Long-term operation	Limited financial budget
	High spatio-temporal resolution	High number of experimental fields, financial budget
	Diverse sensor data with high accuracy	Agricultural activities on the fields (e.g. field cultivation)
	High availability/reliability	Gateways and backbone network
	High extensibility/scalability	Communication stack and energy resources for multi-hop
	Open interfaces	Software updates during runtime
	Additional sensor data transport	Size and sensor data frequency
Site	(Micro)climate condition monitoring	Possibly no power/internet infrastructure
	Soil condition monitoring	Distance to the sites
	Security	Theft protection
Financial	Low price per unit	Partially expensive sensors without low-cost alternatives
		Deployment and maintenance at the test sites
		Additional expenses due to a lack of infrastructure
	Low production price	Multiple parts and components
Practical	High accessibility	Terrain difficult to access
	Simplified manufacturing process	Use of already existing process
	Housing	Limited financial budget and manpower
	Deployment	Number of sensors and soil moisture sensor at a depth of one meter
Technical	Datasheet conformity	CE, WEEE, RoHS compliance, EN 300 220
	Temporal resolution of sensor values	Limited energy budget
	Low response time	Wireless transmissions more unreliable
	Robust design	Thermal and mechanical issues, crop protection

data), mechanical and thermal properties and the financial framework. Due to the large number of test sites and sensors, as well as the acceptance of the farmers to use such a system, the price for a single wireless sensor node and the associated infrastructure must be as low as possible, without compromising quality. Also, the life cycle of a sensor node needs to be taken into account. It is not only a technical solution to be developed, manufactured and deployed at a farmland as expected. Also the maintenance, deinstallation, recycling and sustainability for this kind of project must be taken into account. These and other requirements, as listed in Table 1, are discussed in detail below in a bottom-up approach. Due to the high spatial resolution, a larger number of sensor nodes must be deployed in the geographically distributed test sites. The infrastructure must also be taken into account in the overall project budget. Sensor data needs to be transported from the test sites to a server and stored in a database for further processing. This data must be made available to all project partners. Therefore, the price for gateways, network communication and data transport over the project time, server infrastructure and maintenance, database and its access should be considered. Since microclimatic conditions of the natural environment will be monitored, whose sensor data is building the basis for further processing and decision-making in the upper system layers, the selected sensors must meet certain standards and tolerances. They have to work reliably under difficult environmental conditions such as a wide temperature and humidity range. There should be no need to recalibrate the sensors within the runtime or to maintain them manually, which is not uncommon for environmental sensors. Most technical requirements were defined in the World Meteorological Organisation (WMO) guidelines [40]. They form a reference regarding the temporal resolution and accuracy of the sensors. Table 2 shows operational measurements, tolerances and instrument performances for selected variables defined in [40]. It specifies requirements such as measurement range, deviation, repetition rate and resolution of measurements. Suitable sensors have to be identified and tested according to these specifications or at least as close as possible to them. The large number of wireless sensor nodes does not allow many manual steps in manufacturing, deployment and maintenance. For example, soldering, cabling and housing are very time-consuming tasks, so they need to be reduced or avoided. This limits the use of finished modules and requires their own design and PCB layout. This type of agricultural project also has fairly common practical problems that need to be solved due to mechanical, thermal or environmental challenges. The devices will operate in unguarded areas and will be on duty during all seasons for at least 5 years (possibly 10 years if the project will be extended). That includes a wide temperature range from $-15\,^{\circ}$C up to $+45\,^{\circ}$C, (extreme) fluctuation in humidity, snow, ice, wind, moisture, direct sunlight, possibly heavy rain, hailstorms, thunderstorms, but also wildlife that can become a threat to the installation. Such environmental conditions in particular have a major impact on battery charging and discharging, chip operation, voltage stability at peak current, power consumption and clock drift. Due to the accessibility and the large number of sensor nodes, it should not be necessary to change a battery during the project operational time. This would be associated with finding and accessing nodes or getting in the way of agricultural activities. In order to achieve this long runtime and a high temporal resolution, the sensor node must operate for short periods of time only and needs to implement methods of energy harvesting. The processor, peripherals and particularly the attached sensors must support ultra low power modes with short response or wakeup times. In order to carry out frequent sensor measurements and data transfers, a low power communication protocol, an efficient power management and low power processing are essential. To avoid data loss, the sensor network must have a certain level of reliability and security. This includes node authentication, communication encryption and intrusion detection mechanisms as well as hardware redundancy, flexible software, intelligent error handling and self-healing mechanism. And finally there are requirements based on the project objectives. In order to implement strategies for agricultural management, the data needs to be available in real time. For this project, the term *real time* means within a few minutes, but not more than 1 h. For applications such as a dashboard with sensor data aggregation, for example, a refresh rate of 10 min is also realistic. The WSN has to be implemented and deployed on DAKIS experimental fields in two pilot regions in Germany (Brandenburg and Bavaria), which differ greatly in their agricultural structure. In the pilot region Brandenburg, small-scale management (patch cropping) and differentiated provision of ecosystem services will be tested. The selected landscape windows in Brandenburg are characterized by low annual rainfall and a

Table 2. Operational measurement uncertainty requirements and instrument performance for selected variables [40]

Variable	Range	Reported resolution	Required measurement uncertainty	Sensor timeconstant	Output averaging time	Achievable measurement uncertainty
Air temperature	−80 °C to 60 °C	0.1 K	0.3 K for ≤−40 °C, 0.1 K for >−40 °C and ≤40 °C 0.3 K for >40 °C	20 s	1 min	0.2 K
Relative humidity	0% to 100%	1%	1%	40 s	1 min	3%
Atmospheric pressure	500 hPa to 1080 hPa	0.1 hPa	0.1 hPa	2 s	1 min	0.15 hPa
Wind speed	0 m/s to 75 m/s	0.5 m/s	0.5 m/s for ≤5 m/s 10% for >5 m/s	Distance constant 2 m to 5 m	2 and/or 10 min	0.5 m/s for ≤5 m/s 10% for >5 m/s
Wind direction	0° to 360°	1°	5°	Damping ratio >0.3	2 and/or 10 min	5°
Precipitation amount (daily)	0 mm to 500 mm	0.1 mm	0.1 mm for ≤5 mm 2% for >5 mm	n/a	n/a	The larger of 5% or 0.1 mm
Sunshine duration (daily)	0 h to 24 h	60 s	0.1 h	20 s	n/a	The larger of 0.1 h or 2%

large-scale agricultural landscape on soils with medium yield potential. The sites mostly have sandy soil with low water storage capacity. The farmland has a high site and geographical heterogeneity. The second DAKIS pilot region is located in the *Lower Bavarian Isar-Inn-Hügelland* and the adjacent flat *Inn Valley*, in the district of Passau, Germany. Due to the relief, yield-determining factors such as water and nutrient availability as well as sun exposure and soil conditions vary in a very small space. In contrast, the agricultural areas of the *Inn Valley* are much more homogeneous and easier to manage. This pilot region is characterized by small-scale agriculture with an average field size of less than 2 hectares.

4 Challenges

The requirements and constraints introduced in the previous section create challenges that must be overcome. The main challenges are discussed below. The energy management is strongly challenged by the requirement to operate the sensor network over a period of up to 10 years with the limited energy budget of the sensor nodes. Strategies for energy harvesting must also be developed. The required high spatial resolution leads to a correspondingly high number of sensor nodes. In view of the limited financial budget, this is challenging. The high spatial resolution also leads to the problem of how sensor nodes can be deployed without agricultural activities and sensor network interfering with one another. For example, there is a risk that sensor nodes could be damaged by agricultural machinery. The required high temporal resolution leads to correspondingly short intervals between the measuring cycles. This means that sensor nodes have short sleep times. In addition to long-term operation, this is another opponent for a good energy management. Since diverse sensor data with the highest possible accuracy is needed, a large number of sensors per sensor node is required. This has an impact on both the financial budget and the limited energy budget, since sensor measurement methods can make up a potentially large share on the overall power consumption. To ensure high availability of the sensor data, the sensor network must collect and provide the data quickly and at all times. In order to transport data over a longer distance, multi-hop transmission methods must be used if the radio range of individual nodes is not sufficient. As a result, nodes on the path to a gateway must be more often awake in order to forward data, and therefore consuming more power than other nodes. Since the sensor network must have a certain degree of reliability to avoid data loss, the sensor nodes must also be designed for use under difficult environmental conditions. In summary, it can be said that energy and cost management pose the greatest challenges, followed by thermal and mechanical issues due to the environmental influence.

5 Proposed Approach

In order to design a WSN that is suitable for the agricultural project described in the previous sections, some design decisions need to be made. Various steps are necessary, such as selecting (or developing) suitable sensors, a microprocessor, radio frontends and peripherals, that meet the requirements for measurement precision, performance, energy efficiency and costs. This needs to be merged together into a hardware platform, protected by a housing which withstands the environmental conditions. Conditions in the field and the requirement for spatial resolution have a direct influence on deployment planning, placement of individual sensors and the network coverage and interference. Our proposed solutions to these and other challenges are explained in more detail in this chapter.

5.1 Sensor Selection

Probably the most important components of this WSN are the actual sensors to be able to aggregate data. Before the actual sensor network can be designed, suitable sensors need to be selected or designed. Essential design decisions depend on it. Therefore, the parameters to be monitored have been identified first, followed by research for existing sensors suitable for that task. In an early stage, a pre-selection of sensors has been placed on a sensor test board as shown in Fig. 2, in order to test and measure the usability, software complexity, data accuracy, measurement deviation and power consumption in several modes of operation. The calibration of sensors has been done or is still partially being planned. For example, this task is done by using a climate cabinet, wind tunnel and placing the nodes near to a certified and calibrated Deutscher Wetterdienst (DWD) station with access to its data. Some of these sensor values will probably drift after a currently unknown operating time. Therefore, a strategy for annual repetition to repeat the tests mentioned above with individual sensors from the field has been developed. Deviations can then be compensated, if they are equal for all sensors of the same type. First, the parameters to be monitored by the

Table 3. Preselected sensors for evaluation

Parameter	Sensor	Adjusted range	Precision	Resolution
Air temperature	SHT35	−40 °C to 125 ° C	±0.1 °C	0.01
	BME280	−40 °C to 85 °C	±1.5 °C	0.01
	HS3001	−40 °C to 125 °C	±0.5 °C	0.015
Air humidity	SHT35	0% to 100%	±1.5% RH	0.01
	BME280	0% to 100%	±3.0% RH	0.008
	HS3001	0% to 100%	±3.0% RH	0.01
Atmospheric pressure	LPS25HBTR	260 hPa to 1260 hPa	±0.2 hPa	0.01
	BME280	300 hPa to 1100 hPa	±1.7 hPa	0.18
Sunshine duration	ADPS-9300	640 nm to 940 nm	-	-
UV index	VEML6070	320 nm to 410 nm	5 μW/cm^2	0.05 mW/cm^2
	GUVA-S12SD	240 nm to 370 nm		
Soil temperature	SMT100	−50 °C to 50 °C	±0.2 °C	0.01
Soil moisture	SMT100	0% to 100%	±3.0% RH	0.1

WSN have been identified so that the site-specific properties can be monitored. To monitor microclimatic and soil conditions, the WSN has to carry out sensor measurements for meteorological and soil parameters. There are many questions in DAKIS project intended to be also partly answered based on these sensor values. Therefore, most sensor nodes are equipped with the full range of sensor functions, even if they are only a few meters apart. The following parameters were selected: air temperature and humidity, atmospheric pressure, UV index, sunshine duration, precipitation, wind speed, wind direction, soil temperature and moisture. Precipitation, wind speed and wind direction will be measured by three devices per test site, evenly distributed around the edges of each site. Determining and selecting suitable sensors based on the parameters selected above that fit the WMO requirements is impossible due to the limited budget and the high amount of sensors. Sensors that meet these requirements are very expensive, the field site and environmental conditions are known, so the measurement range can be adjusted. Otherwise realizing this project would not be possible. For example, there will be no −80 °C in Brandenburg, Germany. Even −40 °C is extremely unlikely, but the assumed measuring range increases the possible selection enormously and drastically lowers the price per sensor. Finally a subset of sensors listed in Table 3 have been selected that meet the most requirements, according to Table 2 or to get as close as possible to them. In an early stage, these pre-selected sensors were placed on a specially developed sensor board for further testing, evaluation and calibration as shown in Fig. 2. Using an existing sensor node, it was connected directly to a computer in order to eliminate side effects, trouble shooting issues of irrelevant components and concentrating on usability, configuration options and power measurement in different sleep modes. Laboratory and field site tests have been carried out, e.g. for calibration of meteorological sensors against weather stations matching WMO standards and precision. For calibration, the sensor nodes will be installed/placed near existing DWD weather stations. At the end of this stage, suitable sensors can be selected for integration on sensor nodes. The advantage is that the implementation of drivers, execution of tests and evaluation can be done in parallel with the sensor node development in order to save time and reduce the chance of failures. In the second stage, the collection and processing of sensor data is done on a Microcontroller Unit (MCU)-based sensor node, that has to be developed as well (Table 3).

Fig. 2. Sensor development board with pre-selected sensors

5.2 Sensor Node Concept

The essential components of a sensor node consist of a processing unit, memory, a communication device, sensors and power source that are also part of this node. Beside the pure task of building an adequate sensor node, the reusability is also a major requirement. Therefore, most components should be build as a kind of reusable module in order to save money and time, as well as to increase software quality in future projects. Basically there are two different types of sensor nodes at each test site. There are a total of about 100 Basic Sensor Nodes (BSNs) which are distributed in a fine granular arrangement on the test sites, responsible for collecting most of the sensor data. There are also two to three Extended Sensor Nodes (ExtSNs) per test site, which in addition also record weather data such as wind speed, direction and precipitation. They also act as a gateway, transferring all measurements to a server as well as granting access to the WSN. In relation to the different types, the overall effort on hardware development should be kept as low as possible. There are many processors on the market, having a wide variety of interfaces and peripheral as well as integrated radio frontends. With reusability in mind, the main components have been placed on a core module, named *ATLAS* as shown in Fig. 3. It is including a CC1352R1 MCU with 32-bit Arm Cortex-M4F core and integrated dual band radio for 2.4 GHz and Sub-1 GHz communication, an additional low power 8 MByte NOR flash memory and optional DC/DC converter. All remaining IOs are available through two DC40 pin header, in order to keep this module flexible and increase the reusability of expensive parts. After project completion, the core module can be fully reused in another project. Regardless of their role in the network, this module is used for all sensor nodes. The overall node architecture is illustrated in Fig. 3.

A sufficient selection of the sensors tested in the previous step, will be placed on a board that will form the BSN. It uses the ATLAS core module described above and provides interfaces for programming (cTI-20) and debugging (USB-Serial-Converter). Temperature, Humidity and air pressure will be measured by the HS3001 and LPS25HBTR sensors. As a fallback strategy and for higher precision, the temperature will be also measured in parallel by a NTC. The UV index and sunshine duration will be measured by the VEML6070 and ADPS-3900 sensors. To transfer additional and previously unknown sensor data from partners in the course of the project, an SPI and UART interface as well as 4 GPIOs were reserved. A simple protocol was conceived for the transmission, which is flexible enough by specifying the data payload and its length, for example to transport pictures of photo traps via the WSN. The BSNs will be powered by a 3.2 V 18650 $LiFePo_4$ battery cell, that is charged with a small solar cell as described in detail in the next section.

5.3 Energy Management and Supply Concept

To operate this WSN for more than 5 years (respectively for 10 years), an energy management concept and balanced power supply is required. Using a high power solar cell combined with a car battery is not just oversized and a waste of energy,

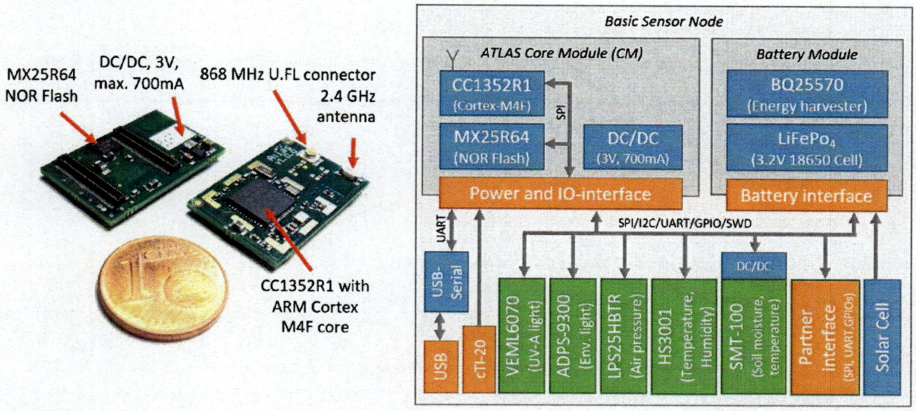

Fig. 3. First version of the core module named *ATLAS* and BSN architecture. **a** *ATLAS* prototype 16 by 25 mm in size. **b** The BSN architecture including the ATLAS core module

it is also a waste of money, resources and increases the risk of theft. An underestimation of the energy consumption inevitably leads to the breakdown of the entire system and generates unnecessary costs for the replacement. Therefore, a small, protected and integrated battery in combination with sufficient energy harvesting methods is required (Table 4).

Table 4. Charging current generated by the energy harvester using different solar cells

Solar cell type	Max. charging current	Completely overcast sky	Min. Avg. daily balance	Max. Avg. daily balance
SLMD960H12	31,7 mA	5,91 mA	38,9 mAh	216,8 mAh
SLMD121H08L	30,2 mA	7,54 mA	37,0 mAh	206,5 mAh
25 × 25 × 3 mm	0,41 mA	0,14 mA	0,5 mAh	2,8 mAh
30 × 30 × 3 mm	10,5 mA	2,05 mA	12,9 mAh	71,8 mAh
2 × 2 25 × 25 × 3 mm	19,0 mA	2,01 mA	23,3 mAh	129,9 mAh

The BSN will be directly supplied by a single 18650 $LiFePo_4$ cell with a capacity of 1600 mAh and a nominal voltage of 3.2 V. This reduces the need for an additional voltage regulator in order to increase the efficiency. Embedded in an aluminum tube and build as a functional module, the battery cell and charger circuit are integrated into a housing concept and can be reused after the project time. All components (except the soil moisture sensor) have been chosen to operate in a range of 2.4 V up to 3.6 V and therefore fit into the $LiFePo_4$ battery cell voltage range of 2.5 V up to 3.6 V. There are no components at the BSN creating a high current flow, so the components voltage range can theoretically be used until the battery has reached its final discharge voltage without the risk of a Brown Out Reset (BOR). In order to protect the battery

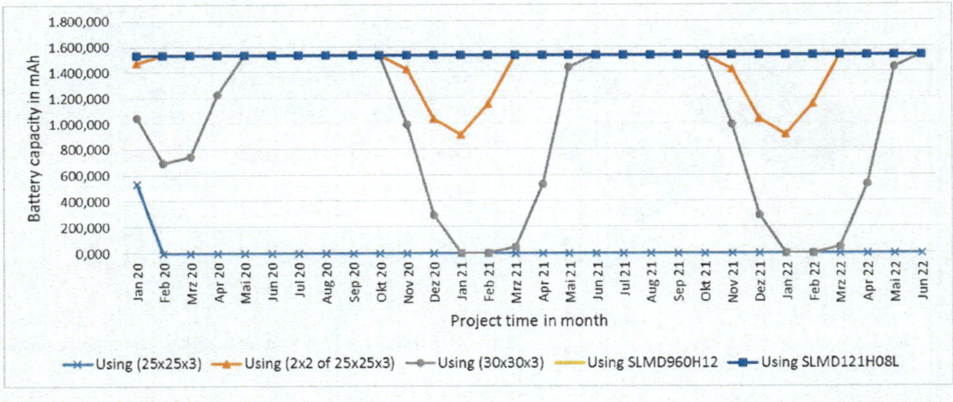

Fig. 4. Battery capacity simulation during the project time, based on DWD weather station and solar cell measurements

and reduce the damaging effects, it will be charged up to 3.55 V and discharged to no less than 2.75 V. The charging process will be controlled by the BQ25570 energy harvesting chip from Texas Instruments, supporting Maximum Power Point Tracking (MPPT) and including an DC/DC buck converter with up to 110 mA. Three evaluation boards have been created and tested before, using the BQ25505, LTC3105 and BQ25570 and all of them worked as specified. This chip was chosen because of its price-performance ratio and the included voltage regulator. The regulator is not enabled or used in this project, but it has been configured to run at 3 V and is intended to be used as fallback strategy and for further projects.

The average power consumption of a single node has been carefully estimated with an upper bound of 1.503 mAh and a peak current of 32 mA (based on datasheet information). Using the data of the average sunshine duration from 1981 to 2010 of publicly accessible DWD weather stations in Berlin and Brandenburg (Germany)[2], a model has been build in order to calculate the daily and monthly battery capacity over the project time. An *hour of sun* is defined as the period of time where the direct solar radiation perpendicular to the direction of the sun is at least 120 W/m^2 [40]. For this, 5 different small solar cells were tested in a experimental way and the charging current has been measured under direct sun light conditions (sun is at its zenith) at the end of June, as listed in Fig. 4. Also measurements with veil clouds and completely overcast sky have been done using our first prototype (Fig. 5). The charging currents have dropped as expected in some cases. However, this also allowed a realistic average to be measured. Based on this measurements, a self discharge current of 100 μA and the DWD data model, the battery capacity during the project time has been calculated as shown in Fig. 4. This estimation is about the *hours of sun* only in order to get an estimated lower bound of the possible charging current based on the DWD data. It is not including the solar radiation below 120 W/m^2, which is used as energy reserve and fallback strategy. With a fully charged battery, the node can operate for up to 44 days, without being charged. Including the daily balance based on the minimum sun hours in mind, the node can operate for

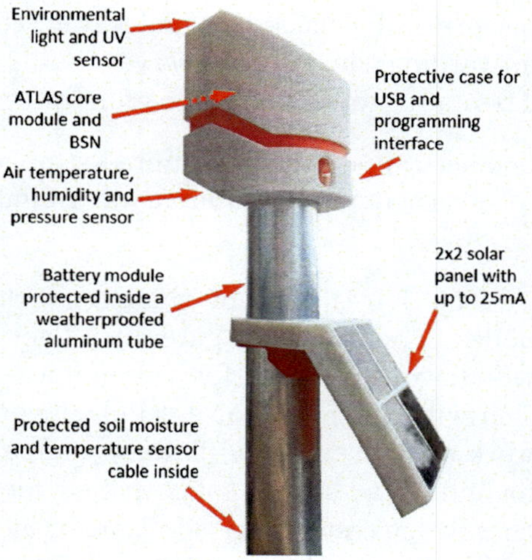

Fig. 5. A first prototype of the BSN with the selected solar cell solution

[2] https://www.dwd.de/DE/leistungen/klimadatendeutschland/mittelwerte/sonne_8110_fest_html.html.

up to 125 days. In this project, we are pursuing a cross-season charging strategy. During the winter season, the average sunshine duration per day drops to a minimum of 1.23 h. This is not enough time to charge the battery or balance its daily total power consumption, so the battery tends to be slowly discharged. In summer it is the other way around with up to 6.48 h and the battery charge will increase slowly day by day.

5.4 Data Management

The essential elements of data management are the sensor network topology in the field, the gateway nodes, the backbone network and the server infrastructure, which will be described in more detail below. In addition to the data acquisition performed by the sensor nodes, the sensor network has the task of collecting and forwarding the data to a server that is collecting all measurements in a database. There are two main reasons for this:

- Sensor nodes have a limited storage capacity
- Sensor data should be made available to the DAKIS DSS in real-time if possible

This job is done by the extended nodes, operating as gateway. They will have a modem for communication with the given local cellular network and an extended battery source including a larger solar panel for energy harvesting. In addition, they will be equipped with a GPS receiver, in order to check or calculate the position by Received Signal Strength Indicator (RSSI) values of all nodes from time to time as an theft information strategy. For communication between the nodes and for internal issues, a mesh topology is intended, as shown in Fig. 6, in which all nodes cooperate to forward the sensor data to one of the gateway nodes. The advantage of this type of topology is that fewer gateways are required because larger distances are covered by multi-hop communication. However, since this type of topology requires nodes to be awake in order to be able to forward sensor data from other nodes, energy-aware routing is essential, including the adjustment of transmission power to save energy. If the negative effects of data forwarding on the energy balance of the sensor nodes are so clear that the planned operating time of the WSN cannot be met, the network can be configured to use a classical star topology with one hop only, which is part of the fallback strategy. Thanks to the dual band radio, this configuration will

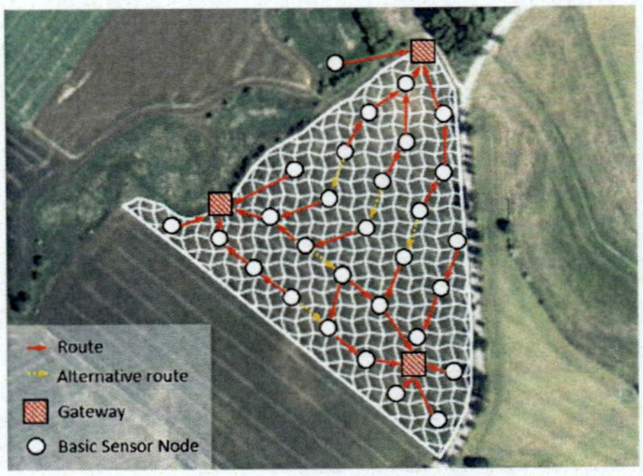

Fig. 6. Intended network topology at one of the test sites in Brandenburg

be possible even after deployment. No energy-aware routing would be necessary since the sensor nodes communicate directly with the nearest gateway. With this option in mind, the gateways will be placed to reach all nodes at a test site directly using the Sub-1 GHz band. The transmission protocols can be changed according to the distances and topology at runtime using an over-the-air update function. In order to transfer the sensor data of the WSNs from the experimental sites to the DAKIS Cloud, a corresponding backbone network must be planned and established, as shown in Fig. 7. Since it cannot be assumed that there is a cable based internet infrastructure in the vicinity of the test sites, the gateway nodes will use a mobile broadband modem in order to be able to transmit data via 2G, 3G, LTE or 5G networks. Furthermore, server infrastructure must be set up to collect the sensor data from the sensor network and make it available to the DAKIS Cloud and project partners. Beside a WSN monitoring function, the configuration and management functionality needs to be implemented.

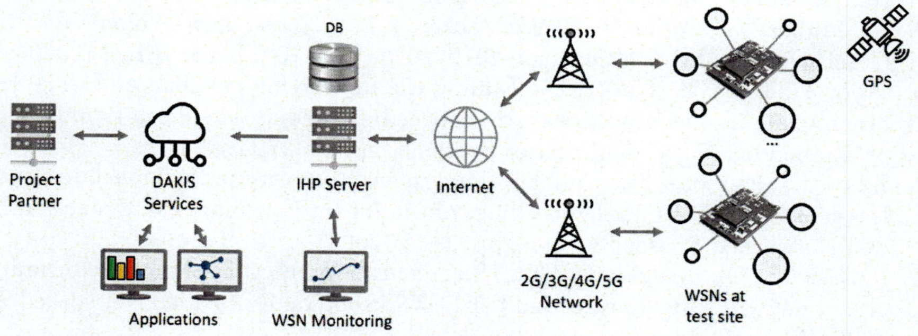

Fig. 7. Backbone network

6 Conclusion

This paper presented challenges in developing a WSN for an agriculture monitoring and decision system in context of the DAKIS project. The proposed approach addresses these challenges and shows solutions and fallback strategies for handling such a project. It describes the current status and the planned steps to develop such a WSN. It can be seen as a blueprint and draws attention to obstacles that scientists may not immediately think of. In addition to the essential challenges such as the financial framework and design decisions, the expansion and modularity to increase reusability were taken into account. Using a practical example, a method for the parallelization of hardware development and restriction of possible sensors was presented. A strategy has been suggested to pre-estimate energy consumption and power balancing based on DWD and geographic data.

References

1. Wolters, V., Isselstein, J., Stützel, H., Ordon, F., von Haaren, C., Schlecht, E., Wesseler, J., Birner, R., von Lützow, M., Brüggemann, N., et al.: Nachhaltige ressourceneffiziente erhöhung der flächenproduktivität: Zukunftsoptionen der deutschen agrarökosystemforschung grundsatzpapier der dfg senatskommission für agrarökosystemforschung. J. für Kulturpflanzen **6**, 225–236 (2014)
2. Collette, L., Hodgkin, T., Kassam, A., Kenmore, P., Lipper, L., Nolte, C., Stamoulis, K., Steduto, P.: Save and grow: a policymaker's guide to the sustainable intensification of smallholder crop production. Food and Agriculture Organization of the United Nations (FAO), Rome (Italy) (2011). http://www.fao.org/3/i2215e/i2215e.pdf
3. Dobermann, A., Nelson, R.: Opportunities and solutions for sustainable food production. Sustainable Development Solutions Network, Paris (2013)
4. Ray, D.K., Mueller, N.D., West, P.C., Foley, J.A.: Yield trends are insufficient to double global crop production by 2050. PloS One **8**(6), e66428 (2013)
5. Bloch, R., Bellingrath-Kimura, S.D.: Smart farming – eine chance für nachhaltige Agrarsysteme? In: Göpel, M., Leitschuh, H., Brunnengräber, A., Ibisch, P., Loske, R., Müller, M., Sommer, J., Weizsäcker, E.U.V. (eds.) Die Ökologie der digitalen Gesellschaft. Jahrbuch Ökologie 2019/2020, pp. 110–116. S. Hirzel (2020)
6. Woo, A., Tong, T., Culler, D.: Taming the underlying challenges of reliable multihop routing in sensor networks. In: Proceedings of the 1st International Conference on Embedded Networked Sensor Systems, pp. 14–27 (2003)
7. Zhao, J., Govindan, R.: Understanding packet delivery performance in dense wireless sensor networks. In: Proceedings of the 1st International Conference on Embedded Networked Sensor Systems, pp. 1–13 (2003)
8. Landsiedel, O., Wehrle, K., Gotz, S.: Accurate prediction of power consumption in sensor networks. In: The Second IEEE Workshop on Embedded Networked Sensors, 2005, EmNetS-II, pp. 37–44 (2005)
9. Ojha, T., Misra, S., Raghuwanshi, N.S.: Wireless sensor networks for agriculture: the state-of-the-art in practice and future challenges. Comput. Electron. Agric. **118**, 66–84 (2015)
10. Jawad, H.M., Nordin, R., Gharghan, S.K., Jawad, A.M., Ismail, M.: Energy-efficient wireless sensor networks for precision agriculture: a review. Sensors **17**(8), 1781 (2017). (Basel, Switzerland)
11. Akyildiz, I.F., Su, W., Sankarasubramaniam, Y., Cayirci, E.: Wireless sensor networks: a survey. Comput. Netw. **38**(4), 393–422 (2002)
12. Baronti, P., Pillai, P., Chook, V.W., Chessa, S., Gotta, A., Hu, Y.F.: Wireless sensor networks: a survey on the state of the art and the 802.15.4 and zigbee standards. Comput. Commun. **30**(7), 1655–1695 (2007)
13. Yick, J., Mukherjee, B., Ghosal, D.: Wireless sensor network survey. Comput. Netw. **52**(12), 2292–2330 (2008)
14. Oliveira, L.M., Rodrigues, J.J.: Wireless sensor networks: a survey on environmental monitoring. JCM **6**(2), 143–151 (2011)
15. Mainetti, L., Patrono, L., Vilei, A.: Evolution of wireless sensor networks towards the internet of things: a survey. In: 2011 19th International Conference on Software, Telecommunications and Computer Networks (SoftCOM), pp. 1–6 (2011)
16. Arampatzis, T., Lygeros, J., Manesis, S.: A survey of applications of wireless sensors and wireless sensor networks. In: Proceedings of the 2005 IEEE International Symposium on, Mediterrean Conference on Control and Automation Intelligent Control, pp. 719–724 (2005)

17. Wang, N., Zhang, N., Wang, M.: Wireless sensors in agriculture and food industry–recent development and future perspective. Comput. Electron. Agric. **50**(1), 1–14 (2006)
18. Li, M., Liu, Y.: Underground structure monitoring with wireless sensor networks. In: Proceedings of the 6th International Conference on Information Processing in Sensor Networks, pp. 69–78 (2007)
19. Ruiz-Garcia, L., Lunadei, L., Barreiro, P., Robla, J.I.: A review of wireless sensor technologies and applications in agriculture and food industry: state of the art and current trends. Sensors **9**(6), 4728–4750 (2009). (Basel, Switzerland)
20. Rawat, P., Singh, K.D., Chaouchi, H., Bonnin, J.M.: Wireless sensor networks: a survey on recent developments and potential synergies. J. Supercomput. **68**(1), 1–48 (2014)
21. ur Rehman, A., Abbasi, A.Z., Islam, N., Shaikh, Z.A.: A review of wireless sensors and networks' applications in agriculture. Comput. Stand. Interfaces **36**(2), 263–270 (2014)
22. Stamenkovic, Z., Randjic, S., Santamaria, I., Pesovic, U., Panic, G., Tanaskovic, S.: Advanced wireless sensor nodes and networks for agricultural applications. In: 2016 24th Telecommunications Forum (TELFOR), pp. 1–8. IEEE, Piscataway (2016)
23. Baggio, A.: Wireless sensor networks in precision agriculture. In: ACM Workshop on Real-World Wireless Sensor Networks (REALWSN 2005), Stockholm, Sweden, vol. 20 (2005)
24. Wark, T., Corke, P., Sikka, P., Klingbeil, L., Guo, Y., Crossman, C., Valencia, P., Swain, D., Bishop-Hurley, G.: Transforming agriculture through pervasive wireless sensor networks. IEEE Pervasive Comput. **6**(2), 50–57 (2007)
25. Langendoen, K., Baggio, A., Visser, O.: Murphy loves potatoes: experiences from a pilot sensor network deployment in precision agriculture. In: 2006 20th International Parallel and Distributed Processing Symposium, IPDPS 2006, p. 8 (2006)
26. Bogena, H.R., Weuthen, A., Rosenbaum, U., Huisman, J.A., Vereecken, H.: Soilnet-a zigbee based soil moisture sensor network. In: AGU Fall Meeting Abstracts (2007)
27. Riquelme, J.L., Soto, F., Suardíaz, J., Sánchez, P., Iborra, A., Vera, J.A.: Wireless sensor networks for precision horticulture in southern Spain. Comput. Electron. Agric. **68**(1), 25–35 (2009)
28. Garcia-Sanchez, A.J., Garcia-Sanchez, F., Garcia-Haro, J.: Wireless sensor network deployment for integrating video-surveillance and data-monitoring in precision agriculture over distributed crops. Comput. Electron. Agric. **75**(2), 288–303 (2011)
29. Srbinovska, M., Gavrovski, C., Dimcev, V., Krkoleva, A., Borozan, V.: Environmental parameters monitoring in precision agriculture using wireless sensor networks. J. Clean. Prod. **88**, 297–307 (2015)
30. Kim, Y., Evans, R.G., Iversen, W.M.: Remote sensing and control of an irrigation system using a distributed wireless sensor network. IEEE Trans. Instrum. Meas. **57**(7), 1379–1387 (2008)
31. Vellidis, G., Tucker, M., Perry, C., Kvien, C., Bednarz, C.: A real-time wireless smart sensor array for scheduling irrigation. Comput. Electron. Agric. **61**(1), 44–50 (2008)
32. Gutiérrez, J., Villa-Medina, J.F., Nieto-Garibay, A., Porta-Gándara, M.Á.: Automated irrigation system using a wireless sensor network and GPRS module. IEEE Trans. Instrum. Meas. **63**(1), 166–176 (2014)
33. Yue, R., Ying, T.: A novel water quality monitoring system based on solar power supply & wireless sensor network. Procedia Environ. Sci. **12**, 265–272 (2012)

34. Lin, M., Wu, Y., Wassell, I.: Wireless sensor network: Water distribution monitoring system. In: 2008 IEEE Radio and Wireless Symposium, pp. 775–778 (2008)
35. Ahonen, T., Virrankoski, R., Elmusrati, M.: Greenhouse monitoring with wireless sensor network. In: IEEE/ASME International Conference on Mechatronic and Embedded Systems and Applications, 2008, MESA 2008, pp. 403–408 (2008)
36. Chaudhary, D.D., Nayse, S.P., Waghmare, L.M.: Application of wireless sensor networks for greenhouse parameter control in precision agriculture. Int. J. Wirel. Mob. Netw. (IJWMN) **3**(1), 140–149 (2011)
37. Malaver, A., Motta, N., Corke, P., Gonzalez, F.: Development and integration of a solar powered unmanned aerial vehicle and a wireless sensor network to monitor greenhouse gases. Sensors **15**(2), 4072–4096 (2015)
38. Burrell, J., Brooke, T., Beckwith, R.: Vineyard computing: sensor networks in agricultural production. IEEE Pervasive Comput. **3**(1), 38–45 (2004)
39. Morais, R., Fernandes, M.A., Matos, S.G., Serôdio, C., Ferreira, P., Reis, M.: A zigbee multi-powered wireless acquisition device for remote sensing applications in precision viticulture. Comput. Electron. Agric. **62**(2), 94–106 (2008)
40. World Meteorological Organization: WMO guide to meteorological instruments and methods of observation: WMO-8 Part I: measurement of meteorological variables, Annex 1.E. World Meteorological Organization, Geneva (2008)

Author Index

B. Ghita and S. Shiaeles (Eds.): INC 2020, LNNS 180, pp. 241–242, 2021.
https://doi.org/10.1007/978-3-030-64758-2